Lecture Notes
in Business Information Processing   552

Series Editors

Wil van der Aalst ⓘ, *RWTH Aachen University, Aachen, Germany*
Sudha Ram ⓘ, *University of Arizona, Tucson, USA*
Michael Rosemann ⓘ, *Queensland University of Technology, Brisbane, Australia*
Clemens Szyperski, *Microsoft Research, Redmond, USA*
Giancarlo Guizzardi ⓘ, *University of Twente, Enschede, The Netherlands*

LNBIP reports state-of-the-art results in areas related to business information systems and industrial application software development – timely, at a high level, and in both printed and electronic form.

The type of material published includes

- Proceedings (published in time for the respective event)
- Postproceedings (consisting of thoroughly revised and/or extended final papers)
- Other edited monographs (such as, for example, project reports or invited volumes)
- Tutorials (coherently integrated collections of lectures given at advanced courses, seminars, schools, etc.)
- Award-winning or exceptional theses

LNBIP is abstracted/indexed in DBLP, EI and Scopus. LNBIP volumes are also submitted for the inclusion in ISI Proceedings.

Roy Jing Yang

Mastering Organizational Dynamics Using Process Mining

Springer

Roy Jing Yang
Queensland University of Technology
Brisbane, QLD, Australia

ISSN 1865-1348   ISSN 1865-1356 (electronic)
Lecture Notes in Business Information Processing
ISBN 978-3-031-93529-9   ISBN 978-3-031-93530-5 (eBook)
https://doi.org/10.1007/978-3-031-93530-5

© The Editor(s) (if applicable) and The Author(s), under exclusive license
to Springer Nature Switzerland AG 2026

This work is subject to copyright. All rights are solely and exclusively licensed by the Publisher, whether the whole or part of the material is concerned, specifically the rights of translation, reprinting, reuse of illustrations, recitation, broadcasting, reproduction on microfilms or in any other physical way, and transmission or information storage and retrieval, electronic adaptation, computer software, or by similar or dissimilar methodology now known or hereafter developed.
The use of general descriptive names, registered names, trademarks, service marks, etc. in this publication does not imply, even in the absence of a specific statement, that such names are exempt from the relevant protective laws and regulations and therefore free for general use.
The publisher, the authors and the editors are safe to assume that the advice and information in this book are believed to be true and accurate at the date of publication. Neither the publisher nor the authors or the editors give a warranty, expressed or implied, with respect to the material contained herein or for any errors or omissions that may have been made. The publisher remains neutral with regard to jurisdictional claims in published maps and institutional affiliations.

This Springer imprint is published by the registered company Springer Nature Switzerland AG
The registered company address is: Gewerbestrasse 11, 6330 Cham, Switzerland

If disposing of this product, please recycle the paper.

Dedicated to my family
In loving memory of my grandmother

Preface

Good leaders know their employees. They strive for effective strategies to fit people together in organizations and fit people's talents with the right jobs. Taizong (598–649 CE), the second emperor and co-founder of the Tang empire, held a firm belief throughout his 23-year-long reign: building an effective workforce was imperative for his governance; and he, as the leader, should keep a good understanding of his clerks and base his appointment decisions on that. Taizong promoted and further developed the imperial examination system so that people were recruited into civil service for their talents rather than lineage or wealth. He also understood that to ensure that his empire thrived, it would take not only the building of a talented workforce but also to review their performance constantly and adjust duties and appointments accordingly. For this purpose, Taizong was open to accepting different voices, with some being harsh critics of his decisions — in the hope that, by doing so, he would get the most accurate information on his staff, especially the ones holding posts remote to the capital. He asked to have the *ping feng* (freestanding privacy screens) in his chamber written with the names of officials along with their achievements and failures[1], so he would be able to analyze and reflect on his decisions on a daily basis. Emperor Taizong's appreciation for talents and the objectivity of information was crucial to successfully leading the state and people from the turbulent early years — faced with the aftermath of war and coups, widespread famine, and risks of domestic rebellions and foreign incursions — to what was later recognized as the golden, prosperous era of Zhenguan. It is also regarded as the pillar of his art of governance, which continued beyond his reign and started a legacy honored by his successors and considered to be a universal model of good governance by historians.

In many ways, managing a modern organization does not resemble ruling an imperial state. However, leaders today would agree with the Tang emperor on the value of accurate workforce insights for effective decision-making. In an era of technological innovation and globalization, modern organizations have no shortage of different forms of data capturing various aspects of business operations. Instead

---

[1] This is recorded in Chapter 197 (Biography 122) in the *New Book of Tang*, a work of history covering the Tang dynasty, compiled by a team of scholars in 1044–1060 CE.

of being underfed, leaders of modern organizations are rather overwhelmed by data. So they would certainly welcome a *ping feng* of their own — not necessarily a screen in a bedroom, but a tool that will enable them to navigate through the overwhelming amount of information and grasp the essential knowledge of their organizations and employees.

This book is written based on the PhD thesis of Jing Yang, which was submitted to the Queensland University of Technology for review in June 2023 and passed examination (defended) in October 2023. It presents doctoral research in the field of process mining, with a focus on developing data-driven methods to discover insights about human resources and their groups in an organizational business process context. This book is suitable for researchers and practitioners in the fields of business process management (BPM) and process mining. It presents an overview of the topic of mining organizational models from event logs and introduces a set of novel ideas, frameworks, and approaches proposed to enhance the state-of-the-art. It also provides pointers to adjacent topics concerned with the organizational perspective of process mining. In the meantime, this book is accessible for people in human resource management (HRM), showing why and how process execution data can be a promising source for workforce analytics.

Book Structure

Chapter 1 provides the general background of the doctoral research, followed by an introduction of process mining as a powerful tool to unlock insights into human resources. It then presents an overview of the related literature in process mining research and capabilities of current process mining software tools. Chapter 2 introduces a conceptual framework formalizing a novel notion of organizational model and the tasks and methods proposed for mining knowledge about resource groupings from event logs. Chapters 3 and 4 present the concrete approaches to discovering organizational models to model resource groupings. Chapter 5 shows a method for applying the framework and model discovery method to support workforce analytics using event logs. Finally, Chapter 6 concludes the book and suggests possibilities for future work based on this research.

Contributions

Our research makes several original contributions to the body of knowledge as follows.

1. We define a new, rich notion of organizational models as the foundation of a novel framework for organizational model mining from event logs. The new organizational models consider multiple dimensions of process execution and link the relevant execution information with resource groupings. As such, they

can be used to capture comprehensive knowledge about resource groups and their involvement in processes.
2. We propose an approach to discovering organizational models from event logs. It is capable of automatically constructing high-quality organizational models from event logs with a minimum set of standard attributes. We present a set of alternative methods for each step of the approach and discuss their configuration.
3. We propose two model evaluation measures: fitness and precision. These measures form a generic basis for assessing the quality of discovered organizational models.
4. We propose a set of model analysis measures and extend them to formulate the notion of resource group work profiles. Work profiles can be extracted from event logs and be used to systematically analyze how resource groups and their members work in process execution, from various aspects and across different periods.
5. The last, but not the least contribution, is made to the research on workforce analytics in the field of human resource management. We introduce business process execution data stored in event logs as a potential data source for analyses, and our approaches contribute a means for workforce analytics to adopt the use of this data source.

Brisbane, Australia *Roy Jing Yang*
March 2025

Acknowledgements

During the past few years, I had conversations with people where the many challenges in completing a PhD project were often mentioned. Sometimes those talks were to "share the burden," and at other times they were about cheering me up. Now, when I look back on the journey, I feel that my memories of those uneasy times seem to have faded away. Part of the reason is that I tend to remember more about the joys of life. In the meantime, I believe it is also because of the luxury I have had on this journey, which supported me to overcome all the uncertainties.

I owe a lot of gratitude to Chun Ouyang. As my principal supervisor, Chun's continued support for me started before I embarked on my trip. She kindly offered key advice during my work on my master's degree (despite having no need or responsibility to do so), which led to a successful collaboration and later my opportunity to study for a PhD. Completing a PhD project certainly involves many challenges, and it is hard to imagine the challenges when it comes to guiding someone through a PhD. Thank you, Chun, for all your help, efforts, and patience (and the mangoes and chocolate as well)!

I am deeply grateful for the support from Arthur ter Hofstede. Arthur has been an "all-around" advisor. From Arthur, I sought advice on various aspects — how to write good formalization, how to better express myself, how to collaborate with people effectively, etc. — and Arthur always offered thoughtful feedback that I could take to develop my capabilities as a researcher and beyond. Thank you, Arthur.

Particular gratitude also goes to Wil van der Aalst. I still remember the day when I asked Wil stiffly whether he could be on my supervisory team. Both Wil and I were visiting QUT, only that Wil was an established scholar who pilots research in the field, while I was working on my master's degree and had not secured any PhD scholarship; and Wil hardly knew anything about me! Thank you, Wil, for accepting that invitation, and for all the insightful advice and feedback that you have provided to me.

I also thank Yang Yu. His support and guidance during my bachelor's and master's studies were key to the start of my journey of research. And I always find encouragement when talking with him.

I would also like to thank many people who have helped me in many ways. Michael provided key input into some of the co-authored work related to my thesis. Guy, Karen, Yue, Catarina, Renuka, Erwin, Barbara, Pnina, Wasana, Robert, Alistair, Sander, Moe, Colin, and David offered comments and asked questions that enabled me to reflect on my research from different perspectives. I worked with Paul, Guvenc, Miranda, Amy, Hamish, Belinda, and Nigel on other projects during the course of my PhD. From them, I have learned a lot. Special thanks go to my fellow students: Adam Banham, Adam Burke, Anu, Atae, Azumah, Behnam, Bemali, Chester, Christoph, Felipe, Ina, Jenny, Joe, Kenny, Lakmali, Lauren, Leon, Malmi, Miguel, Mostafa, Mythreyi, Pamela, Richard, Sareh, Tendai, Yancong, Zippo — for sharing the joys and burdens of doing a PhD. I also thank people who offered assistance to me and whom I shared conversations with.

Last and most importantly, many thanks to my family. I would never be able to come this far without their unwavering support, especially when being thousands of kilometers away from home and at a time of uncertainty. I am most indebted to my wife, Wenhui, for all her love and encouragement, which have always been the pillar of my adventures, past and forward.

Contents

1 **Introduction** 1
  1.1 Process Mining 2
  1.2 Mining Organizational Models from Event Logs 7
  1.3 Outlook 13

2 **Framework for Organizational Model Mining** 17
  2.1 Preliminaries 17
  2.2 Execution Context 21
  2.3 Organizational Model 23
  2.4 Discovering Organizational Models 25
  2.5 Evaluating Organizational Models 26
    2.5.1 Fitness 26
    2.5.2 Precision 27
  2.6 Analyzing Organizational Models 28
  2.7 Discussion 30

3 **Learning Execution Contexts** 33
  3.1 Preliminaries 34
  3.2 Problem Modeling 35
    3.2.1 Categorization Rules 35
    3.2.2 Quality Measures for Execution Contexts 36
    3.2.3 Problem Statement 38
  3.3 Problem Solution 39
    3.3.1 Deriving Attribute Specification 39
    3.3.2 Inducing Rules via Simulated Annealing 44
  3.4 Evaluation 49
    3.4.1 Event Log Datasets 49
    3.4.2 Experiment Setup 53
    3.4.3 Evaluation against the Baselines 55
    3.4.4 Evaluation between tree-based and SA-based 56
    3.4.5 Summary 60

	3.5 Discussion 60
4	**Discovering Organizational Models** 63
	4.1 A Three-Phased Discovery Approach 63
	4.1.1 Determining Execution Contexts 64
	4.1.2 Discovering Resource Grouping 65
	4.1.3 Profiling Resource Groups............................. 67
	4.2 Implementation 68
	4.3 Evaluation 68
	4.3.1 Experiment Setup 69
	4.3.2 Model Evaluation and Comparison 71
	4.3.3 Model Diagnosis...................................... 74
	4.3.4 Summary... 77
	4.4 Discussion 78
5	**Applying Organizational Models to Workforce Analytics** 79
	5.1 Preliminaries 79
	5.2 Resource Group Work Profiles................................ 81
	5.2.1 Work Profile Indicators 81
	5.2.2 Extracting and Analyzing Work Profiles 82
	5.3 Case Study: One Process, Five Municipalities 89
	5.3.1 Group-level Analysis 90
	5.3.2 Within-Group Analysis 92
	5.3.3 Summary... 94
	5.4 Discussion 95
6	**Epilogue** 99
	6.1 Conclusions 99
	6.2 Future Work 100
	References 103

Chapter 1
Introduction

> *A wise leader knows how to fit the right personnel with the right tasks, like a skillful carpenter knows how to utilize timber of any shapes or lengths.*
>
> Taizong of the Tang empire

Many leading enterprises have started seeking opportunities to leverage advanced analytics on employee-related data to provide evidence-based insights into their workforce [65]. Among others, Google's project "Oxygen" is an example of successfully deploying workforce analytics, which helped improve the company's productivity and employee well-being and build up effective human resource management practices [40]. Yet, there are practical challenges that prevent workforce analytics from realizing its promise. One notable challenge is concerned with the absence of group-oriented analysis pivotal to strategy execution and organizational effectiveness [56]. For example, current workforce analytics has not yet enabled consistent comparisons across internal groups within organizations [44].

In modern organizations, employees, along with other resources, are deployed in business processes [37] to deliver products and services. To maintain competitiveness in an ever-changing environment, organizations have to be able to rapidly adapt their business processes and optimally marshal their resources. Often, business processes are "end-to-end", i.e., they cut across organizational boundaries and collectively involve human resources from different functional units, linking the performance of employees and their organizational groups with process outcomes. Therefore, organizations need to possess the capability to constantly evolve organizational groups alongside changing business processes [29], and it is thus imperative that they maintain accurate and timely insights into the groups [30]. Clearly, relying on organizational charts — too static and often too high-level — or on leaders' intuition — too vague and often anecdotal — will not be conducive to achieving this capability.

Process execution data provides a promising source for extracting accurate and timely insights into human resources in the context of business processes [68]. This data is readily available in many contemporary "process-aware" information systems [7] and is often stored in so-called event logs. Event logs record activities undertaken at a specific time in the context of the execution of a certain instance of a process (often known as a case) [7, 37]. In addition, they may record resources who executed those activities. As such, event logs capture the trails of human resource participation in various contexts of actual business process execution. This makes event logs useful input to workforce analytics, complementing contemporary HR data sources such as annual employee surveys and competency profiles. Effective use of event logs has the potential to addresses one of the key questions in HR management [56]: How may the vast amount of enterprise IT system data be exploited for analyzing the connection between employee behavior and operational performance of the organization?

In the remainder of this book, we explore *process mining* — a body of principles, methods, and techniques for the systematic analysis of event logs — as a powerful tool to enable workforce analytics in the context of business processes.

1.1 Process Mining

Process mining extracts knowledge about processes from data recording process execution. The extracted knowledge provides insights into process management and improvement [5, 7, 31]. Process execution data is already recorded [7] in many contemporary enterprise information systems such as Workflow Management (WfM) systems, Enterprise Resource Planning (ERP) systems, and Customer Relationship Management (CRM) systems. These information systems are process-aware [36], i.e., they support the deployment of business processes in organizations captured in an explicit notion, and they are involved in the management of that process (not necessarily controlling the process like a workflow engine dispatching work to employees). As such, data extracted from these systems can be used to provide information on how processes execute *in reality*. On the contrary, modeled processes, which are often used at the initial design phase of a business process, do not reflect up-to-date changes once the process has been deployed.

Process mining is a growing discipline that researches the principles and methods to extract insights from event logs for discovering, monitoring, and improving processes in organizations. As illustrated in Figure 1.1, process mining bridges data science with process science, creating synergies between emerging data-centric analysis techniques and tools and traditional model-focused analysis [7]. It emphasizes gaining insights into the current-state process, for example, performance bottlenecks and unexpected variation, and creates a transparent view of complex organizational activities through exploiting warehoused data. Therefore, process mining is considered a promising solution to some of the most fundamental challenges in business process management [31].

The value of process mining can be further explained by considering its contributions to the lifecycle of process management [37]. Figure 1.2 illustrates the idea: Organizations use process models to specify how their business processes operate, incorporating employees and other core assets. Process models are then used to configure process-aware information systems, which are deployed to help organizations implement and control their business processes. These process-aware information systems record the "trail" data of actual process execution, which can be extracted into the form of event logs. Process mining provides a means to utilize event logs for

1.1 Process Mining

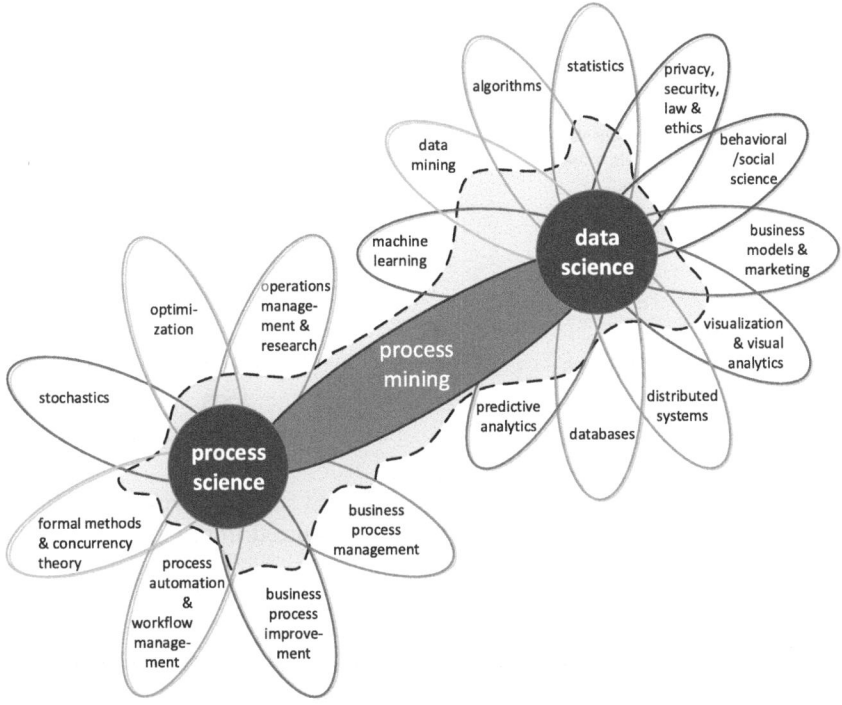

Fig. 1.1: Process mining bridges data science with process science (this figure is sourced from Figure 1.7 in [7]).

deriving knowledge about the process, which can be used together with the initial process model to support the improvement of processes. A recent study [2] predicts that the market for process mining will grow tenfold over the next few years.

Process mining can unearth insights into processes, but is not limited to the discovery of process models that describe process activities and their ordering (i.e., the "control-flow" of a process). Many types of process mining exist, including: (i) *discovery*, which aims to construct a model from a given event log to represent the process in reality as recorded in the data; (ii) *conformance*, which aims to compare a process model that describes a process against the observations in an event log, e.g., investigating if there are compliance issues in the execution of a process; and (iii) *enhancement*, which aims to extend or improve a process model by using event log information, e.g., annotating a model with case and time information to show the performance bottlenecks in a process.

In the pasts decades, many process mining software tools have emerged. A curated list of current tools, including open-source and commercial ones, is available at the process mining website[2]. Readers are also referred to the process mining software

---

[2] http://www.processmining.org/

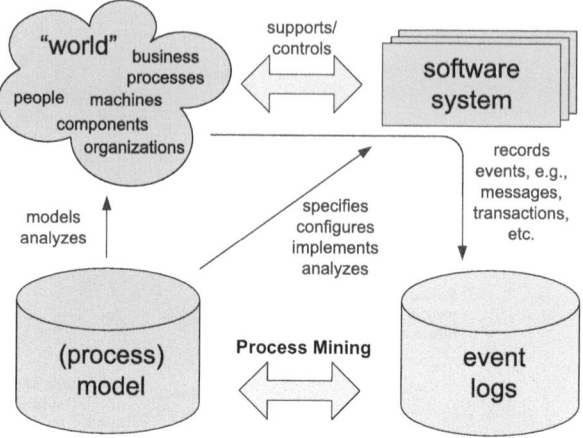

Fig. 1.2: Process mining can provide evidence-based support to key phases in the management of business processes (this figure is sourced from Figure 2.5 in [7], with adaptations).

comparison website[3] for a systematic survey of the capabilities of existing process mining tools [85]. Many tools provide academic license for a learning or research purpose. Some are aimed for deployment at an enterprise level and are equipped with capabilities to be connected with existing enterprise information systems (such as SAP and Workday) [8] and allow for continual analysis of process data concerned with large number of stakeholders in an organization. Process mining already has found applications across multiple industries, e.g., finance, logistics, healthcare. The Process Mining Handbook [8] presents a number of real-world use cases of process mining technology in healthcare and auditing contexts.

While lion's share of research has devoted to studying the process control-flow, process mining is also concerned with various other perspectives, including: (i) the *organizational/resource* perspective, which looks into how (human) resources *participated* in process execution and how they are *related* in that context; (ii) the *case* perspective, which focuses on the properties of cases, e.g., analyzing how the characteristics of cases may be related to the control-flow paths or other data elements; (iii) the *time* perspective, which investigates the temporal aspect of a process, e.g., predicting the remaining processing time of running cases.

This book is concerned with the organizational perspective of process mining, also known as *organizational mining* [77]. Many event logs record resource identifiers, i.e., the identity of (human) resources who triggered the events, in addition to the minimally required data (cases, activities, timestamps). Some event logs may record additional resource information such as roles or organizational groups. In the IEEE XES standard [4] for event logs, they are captured by event attributes of

---

[3] https://www.processmining-software.com/

1.1 Process Mining

the organizational extension [3]. However, it is often assumed that only resource identifiers are present in a log (i.e., the role and organizational group of resources are unknown). This is the starting point for many process mining methods that aim at extracting insights from an organizational perspective.

Van der Aalst and Song [10] first suggest that two types of resource-related knowledge can be discovered using event logs, i.e., organizational structures revealing the possible business roles of individuals and resource social networks describing inter-resource relationships in process execution. Their subsequent work [9] focuses on *social network mining* and identifies four types of resource relationships that are discoverable: the handover of work between resources, similarities in performing activities ("joint activities"), simultaneous appearance in cases ("joint cases"), and relationships determined by special types of events (e.g., delegations of work). Discovered networks consist of nodes representing individual resources and links representing the strength of some relationship. Then, existing social network analysis techniques can be applied to the discovered networks for identifying frequent interactions, modeling information flows, etc. A case study reported in their work shows the usefulness of mining social networks and suggests the potential values of mining other resource-related insights from event logs. Further studies on the topic of social network mining explore the applications of various network analysis techniques and the insights they provide. For example, Ferreira and Alves [39] propose to apply community detection on resource social networks to facilitate the analysis and visualization at different levels of abstraction; Liu et al. [58] propose to discover features like social positions from resource social networks and use them to augment resource models for supporting team collaboration in task assignment; Kumar and Liu [55] utilize the handover ("hand-off") patterns discovered from event logs to gain insights into frequent interaction patterns of resources and their impact on process performance.

Ly et al. [62] propose the problem of *assignment rule mining*, which aims at extracting inherent rules that decide the assignment of tasks (process activities) to resources. The input to assignment rule mining consists of an event log and a given organizational model. A decision tree learning technique is employed and adapted for rule extraction. Discovered staff assignment rules can help process owners and system engineers diagnose and refine the assignment mining [69, 60, 48, 59, 73] varies in terms of the required event information and the techniques applied.

Resource behavioral profile mining aims at characterizing different aspects of individual resource behavior in process execution. Pika et al. [68] define several aspects of resource behavior that can be measured and analyzed given event logs, e.g., resource skills, utilization, and preference. Mining resource behavioral profiles can provide objective information to analysts regarding resource performance in the context of process execution. Other research on the topic considers different types of resource behavior, e.g., Nakatumba and van der Aalst [66] propose to discover from event logs the relationship between resource workload and resource efficiency at work; Huang et al. [49] introduce the characterization of resource availability and competence using event logs; Suriadi et al. [81] focus on discovering the work prioritization patterns of resources, their conformance to prescribed discipline of queuing, and how such prioritization patterns impact the overall process performance.

Organizational model mining aims to discover the organizational structures deployed around resources. As event logs may capture only fractions of employees' activities in organizations [77], mining organizational models is often formulated as identifying groups of resources exhibiting similar characteristics in process execution. The majority of the state-of-the-art research addresses the problem by first characterizing how individual resources participate in process execution or how they interact with each other. Then, the problem is transformed into a clustering (e.g., [51, 88]) or a community-detection problem (e.g., [13, 89]), and dedicated techniques are applied to solve them. As a result, a discovered organizational model comprises groups of resources with shared features, e.g., frequently performing a unique subset of process activities. These models may offer insights into resource planning and the design of business roles.

Despite the different forms of discovered organizational knowledge, the above research topics may be related. For example, the discovery of resource groups in organizational model mining can be achieved by applying graph partitioning or community detection techniques on resource social networks extracted from event logs [77, 39, 13]. It is also worthwhile mentioning that there exist different views toward the resource-oriented process mining literature. Schönig et al. [73] discuss existing mining methods based on their support for discovering resource patterns [71] from event logs. Zhao and Zhao [92] consider role discovery separate from organizational model mining and place mining of resource assignment rules and resource behavior together under the term of resource allocation.

Current process mining software provides some support for analysis from the organizational/resource perspective. For example, PM4Py [18], an open-source Python software library for process mining, implements several algorithms for social network mining [9] and organizational model mining (as business role discovery [26]); IBM Process Mining[4] offers support for resource profile mining by reporting resource workload statistics over time and for social network mining by displaying clustering resources with frequent interaction in process execution (see Figure 1.3); Microsoft Power Automate[5] provides a "hierarchical process mining" feature to enable displaying resource grouping information (e.g., cost centers) on top of groups of process activities in a process map, which can be considered as process enhancement with organizational information.

In addition, many leading process mining software tools [1] offer capabilities of analyzing human resource participation in a process in the form of *task mining*. For example, Celonis[6] offers a "workforce productivity" analysis module to visualize desktop application usage and copy-paste behavior at a resource-group level, which assists with identifying best practices for employees and opportunities for task au-

---

[4] https://www.ibm.com/products/process-mining
[5] https://www.microsoft.com/en-us/power-platform/products/power-automate
[6] https://www.celonis.com/solutions/workforce-productivity/

1.2 Mining Organizational Models from Event Logs

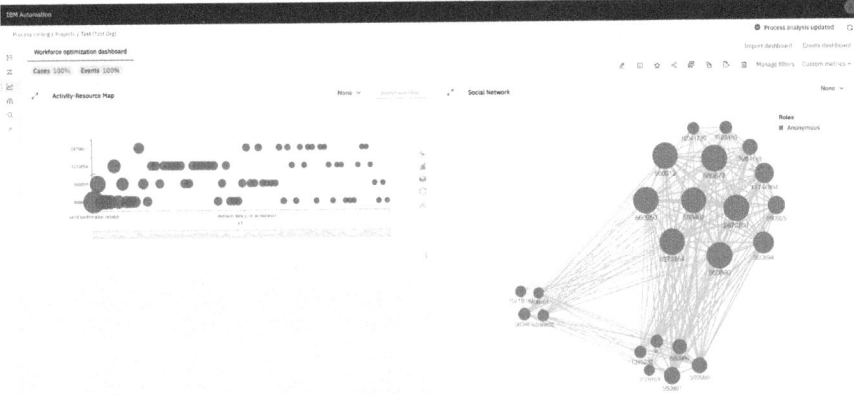

Fig. 1.3: IBM Process Mining, one of the current process mining software, provides support for resource profile mining (Left: visualizing resource workload on different types of activity over time) and social network mining (Right: visualizing resource interactions).

tomation; Pega[7] offers a "workforce intelligence" solution, which reports on time distribution and usage pattern of desktop applications and recommends automation opportunities to reduce costs. UiPath[8] and Microsoft Power Automate also offer task mining features alongside their Robotic Process Automation products. While task mining and process mining may be considered complementary technologies, it is worthwhile noting the difference: Task mining is conducted at a process activity level, using user interaction data collected when human resources are using their desktops or laptops to perform allocated tasks; process mining is conducted at an organization-wide process level, using event logs recorded by enterprise information systems.

1.2 Mining Organizational Models from Event Logs

Organizational model mining [77, 13, 88] is a relatively underexplored subfield of process mining. It is concerned with the study of groups of human resources, specifically how models can be derived from event logs to reflect resource groupings in process execution.

---

[7] https://www.pega.com/products/platform/workforce-intelligence
[8] https://www.uipath.com/product/process-mining

Organizational Model Discovery

The main focus of organizational model mining is model discovery, i.e., given an event log recording actual process execution, constructing a descriptive model that reflects the reality captured by the log data.

Some existing research [51, 91, 26, 16] defines an organizational model as a set of business roles. The early work by Jin et al. [51] considers resources performing similar types of tasks in process execution as having the same roles. Based on the idea, a performer-by-activity matrix [9] is first derived using the input event log, and it is then fed to a clustering algorithm to distinguish between resources. The discovered roles in their models lack a clear description, and additional knowledge is required to interpret the meaning of those roles.

Zhao et al. [91] further take into account interactions among resources besides the similarity in task execution to determine the set of roles. They formulate the model discovery problem as an integer optimization problem, under the assumption that the interactions among resources should conform to their role designation. This method was claimed to have better performance [91] but, again, focused merely on distinguishing among resources.

Adopting a similar idea of utilizing resource interactions, Burattin et al. [26] propose a method to identify business roles based on clustering process activities rather than clustering resources. First, given a process model and an event log, edges connecting process activities are filtered based on the handover of work; then activities are partitioned; finally, roles are derived after iterative modification of the activity partition. A discovered model is a partition of process activities, each representing a role performed by a few resources recorded in the log. The grouping of resources is not necessarily disjoint. As such, their approach allows meaningful interpretation of the grouping in a discovered model.

Baumgrass [16] presents an approach to derive the up-to-date role-based access control (RBAC) policies from event log data in the context of role engineering. This objective leads to the discovered organizational models being defined as not just the grouping of resources into roles, but also the "permissions" assigned to those roles. Baumgrass [16] also proposes the possibility of characterizing hierarchies within the models, based on how the permissions of roles are related. The discovery of the models from a given event log is achieved by using a set of pre-defined rules mapping standard event log attributes (in the XES standard [4]) to RBAC artifacts.

Song and van der Aalst [77] recognize that it is challenging to fully recover the "actual organizational model in an organization" due to the limitation of information recorded in event logs. In light of this, they define organizational models as consisting of (i) organizational entities that can be mapped onto various organizational groupings like project teams, organizational units, functional departments, and may be linked to each other in a hierarchical structure, and (ii) relationships between organizational entities and process activities, which represent the capabilities or responsibilities of the groups of resources. The discovery of such models can be approached using cluster analysis or by graph partitioning on resource social networks discovered from event logs. The concept of organizational models specified

1.2 Mining Organizational Models from Event Logs

by Song and van der Aalst is adopted by many subsequent works on organizational model mining, in which different techniques for discovering groups are applied. Ni et al. [67] address the model discovery problem on large-scale event logs by applying a grid-based clustering technique, which has the advantage of a fast processing time independent of the number of resources. Appice [13] and Ye et al. [89] use community detection techniques, and Yang et al. [88] use clustering techniques to derive organizational models where resources are allowed to be members of multiple groups. Their research aims to construct models that can better describe the overlaps between resource groups, which are common in real-life organizations.

The existing research introduced so far exploits event log information on how resources performed different process activities and how they interacted with other resources. Some work on organizational model discovery utilizes other information. Delcoucq et al. [32] considers clustering resources based on how they performed activities in different orders, captured by the so-called local process models [84], i.e., subprocesses describing frequent process control-flow [7] patterns. This idea allows the discovery of more fine-grained resource groupings, compared to the previous approaches [77, 67, 13, 89, 88]. Van Hulzen et al. [50] propose the notion of "activity instance archetype" to capture contextual factors impacting how activity instances were executed. An activity instance archetype consists of (derived) event attributes to enable quality clustering of events. Given an event log, activity instance archetypes can be discovered by applying model-based clustering on events enriched with selected attributes. Then, resources are characterized by their execution of the discovered activity instance archetypes. Finally, resource groupings concerned with contextual factors can be discovered. Note that the contextual factors may include rich data beyond the activity labels in the input log.

Some work on discovering organizational models assumes the presence of context-specific information in input event logs. Li et al. [57] propose a way to discover organizational models in the context of knowledge maintenance organizations. A model is represented as a graph where nodes correspond to resources at some level of the organizational hierarchy and links describe the transfer of knowledge from the more superior staff to the others. Model discovery exploits the interactions among resources and categorizes resources based on the similarities of their interaction patterns. The core idea is comparable to that of Zhao et al. [91], only that Li et al. [57] focus specifically on the context of knowledge transfer, i.e., interactions among resources refer specifically to the handover of organizational knowledge.

In the work by Hanachi et al. [43], an organizational model is defined as a graph where nodes are the resources and links define coordination among resources. Annotations on such a graph indicate specific structural information such as hierarchy and federation [43]. The procedure of constructing such a model relies on extracting interactions of resources and investigating the connectivity of the graph. Compared to other existing research, Hanachi's idea is more related to the notion of organizational structures in organizational theory [29], i.e., modeling both organizational groupings and communication patterns. But since the model discovery requires additional semantic information on the coordination types between resources, the applicability of their approach is subject to the availability of that semantic information in an input log.

A similar idea is applied in the research by Sellami et al. [74] and Bouzguenda and Abdelkafi [22]. Both consider an organizational model as a meta-model enriched with the so-called organizational ontology. This meta-model consists of resources ("performers"), roles, organizational units, and resource membership. It is enriched by a set of ontologies to characterize complex resource relationships such as cooperation and subordination. To discover such models, the input event logs are expected to contain additional information based on the proposed ontology, and the outputs are graphs where nodes may denote any entity in the meta-model (resources, roles, and organizational units) and links denote the relationships implied by the interactions among resources recorded in the logs. Note again that the applicability of these approaches depends on whether an event log records the required ontological information.

Model Discovery: the Research Gaps

We adopt three perspectives when reviewing the process mining literature on the discovery of organizational models: (i) the use of event log information on the multiple process dimensions, (ii) the interpretation of discovered organizational models, and (iii) the evaluation of the quality of the discovered organizational models. Our review below shows that existing methods for organizational model mining are not fully up to the task of supporting analyses of resource groupings.

First, an event log recording process execution typically encompasses multiple dimensions including case, activity, and time. For organizational model mining, an input log need to also include resource identifiers. Hence, the participation of human resources in process execution can be analyzed based on (i) how resources perform activities, (ii) how they are involved in different cases, (iii) how they work at different times, and (iv) how they interact with each other in process execution. With the abundant multidimensional information, model discovery methods should *maximize the use of relevant input event log information necessary* for discovering organizational-grouping-related knowledge. Most existing methods exploit the information on resources performing activities. This is because common resource grouping schemes (e.g., business roles, functional units) often result in specialized groups of employees handling specific activities in a process. Some methods exploit the information on resource interactions (e.g., handover of work between resources executing adjacent process activities), in particular, studies that focus on the reporting relationships among employees [43, 74, 22]. Yet, information related to cases and time is rarely considered when discovering organizational models. This narrow focus is limiting when resource groupings need to be considered across different cases (e.g., specialist groups dedicated to particular customers) or across different time periods (e.g., employees playing the same role but working different shifts). Song and Van der Aalst [77] exploit case information in event logs and discover employee

1.2 Mining Organizational Models from Event Logs

teams assembled for collective tasks. Van Hulzen et al. [50] exploit the "contextual factors" impacting the execution of activities, which may include case and time information. Hence, their work contributes a novel attempt to utilize multidimensional process information.

In the meantime, some existing methods require information additional to event logs, including the transfer of knowledge [57], coordination types [43], and ontological data on resource interaction types [74, 22]. However, such information is not generally available, imposing limits on the applicability of those approaches.

Second, to allow for the interpretation of discovered organizational models, model discovery methods should *ensure that the output models capture the involvement of resource groups in process execution*. In all existing work, discovered organizational models capture the groupings of human resources. But only few [91, 26, 16, 77, 32, 50] allow models to describe how the discovered resource groups are involved in process execution. This limitation makes organizational models in most literature not very helpful in analyzing and understanding the behavior, responsibilities, or permissions of resource groups. Establishing the connection between discovered resource grouping and process execution is important for interpreting discovered organizational models.

Model discovery methods should *allow the output models to be evaluated independently and objectively*. That is, the evaluation of the quality of discovered models uses only information in the input event log, without relying on additional prior knowledge; and the evaluation is not subject to the specific techniques applied in the discovery. We synthesize several means of evaluation in the literature. Some existing studies only demonstrate how their proposed methods may be applied to an event log, either synthetic or collected from real-world process execution, to discover organizational models [16, 91, 67, 43, 74, 22, 32]. But there lacks an indication of the quality of the discovery outputs. Some evaluate the quality of discovered models by comparing them to some domain knowledge relevant to resource groupings in the process, i.e., official organizational structures or business roles [77, 51, 26, 88, 32, 57]. Clearly, this relies on the availability of such domain knowledge. In addition, the evaluation results can be flawed, since human resource groupings in reality may deviate from the domain knowledge used as the reference. Another means is to assess the effectiveness of the techniques applied for model discovery, which often requires using a method that is specific to the techniques. Both Appice [13] and Ye et al. [89] apply community-detection techniques to discover organizational models and adopt modularity measures to evaluate how effectively those community-detection techniques perform. However, modularity measures are specific to community-detection problems and cannot be applied to evaluate models discovered using other techniques, e.g., those based on cluster analysis [77]. In addition, since the problem of discovering organizational models can be considered a type of *knowledge discovery from data*, we consider it necessary to assess the quality of discovered models, i.e., the output knowledge, against the input event log data. Note that this idea is consistent with how model quality is typically evaluated in process mining research [37, 7], i.e., by comparing modeled behavior to log observations. However, we did not find any existing study that considers the use of input event logs for evaluating the quality of discovered organizational models. A generic model evaluation approach is still missing.

To recap, we identify three research gaps that may impede the use of the existing organizational model mining approaches in practice, as illustrated in Figure 1.4.

1. Event log information typically used for organizational model mining records multiple process dimensions, but it has not yet been fully considered for model discovery.
2. Many approaches do not describe the involvement of resource groups in process execution, which in turn, makes it challenging to interpret the discovered resource groupings or use the discovered models for analyzing the behavior of those resource groups.
3. A generic method for evaluating discovered organizational models based on input event logs is yet to be proposed.

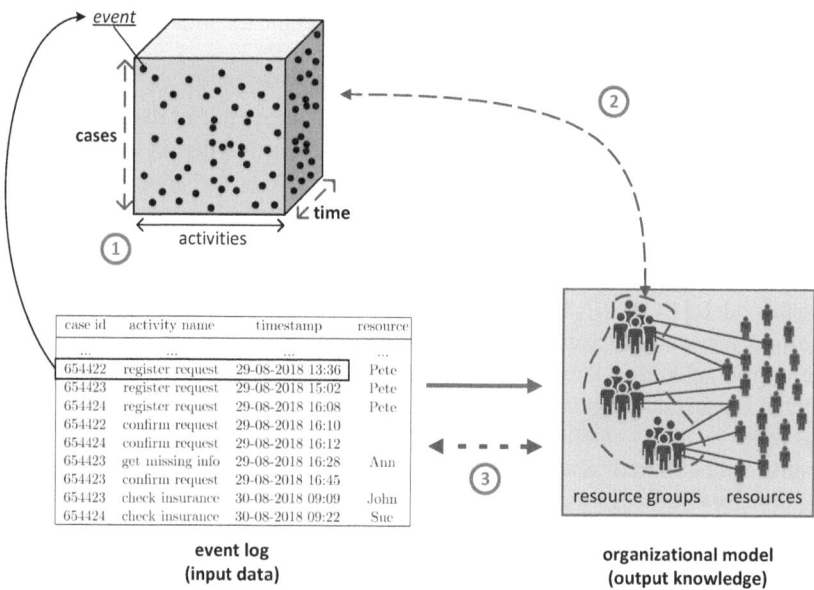

Fig. 1.4: An illustration of the three research gaps identified in existing organizational model mining research: (1) lack of consideration for multiple process dimensions; (2) missing description of discovered resource groups in terms of their involvement in process execution; and (3) absence of evaluation between input event logs (data) and discovered models (knowledge).

Beyong Model Discovery

Some existing research is concerned with resource groupings but does not focus on the discovery of organizational models.

The work by Baumgrass et al. [17] attempts to check the conformance of organizational models against event logs. This work uses the model definition in the previous work by Baumgrass [16], i.e., an organizational model consists of a set of roles and their permissions. In this work, they propose an approach to checking if the process execution as recorded in event logs conforms to given RBAC policies. Given an organizational model representing some RBAC policies, the model is translated into Linear Temporal Logic (LTL) for conformance checking. Comparing the LTL predicates with the event log may reveal violations of the given policies. As such, this work also resembles the idea of assignment rule mining (see Section 1.1) that compares resource-activity relationships extracted from event logs to predefined ones.

There is also research that considers using event log data to extend organizational models. The input consists of an event log and an "as-is" organizational model, and the output is an enhanced model annotated with log information. Model extension considers the use of resource interaction information [77] and temporal information [13, 90].

Organizational models can be enriched by projecting resource interaction information extracted from event logs onto the resource group level. In the work by Song and van der Aalst [77], a method for analyzing the information flows between organizational entities is presented. It aims to uncover the communication patterns of resource groups by aggregating resource interactions (extracted from social network mining, see Section 1.1) and mapping them onto the links between the organizational entities that the resources belong to.

Another way to extend organizational models uses the temporal information stored in event logs. When timestamps of events are recorded in a log, it is possible to slice an event log into several time segments, with each corresponding to a sub-log. Then, organizational model discovery can be applied to each of the sub-logs. This allows tracking the changes of organizational models over time [13] and studying the "lifecycle" of resource groups. In [90], a similar method is proposed, which applies organizational model discovery to event streams to enable online analysis of organizational models. We can view these methods as generating "snapshots" of an organizational model at different times.

1.3 Outlook

This book introduces novel research with an aim to address the aforementioned research gaps in organizational model mining. Specifically, we seek to offer solutions to a couple of key questions as follows. Fig. 1.5 provides an overview of our research and its artifacts.

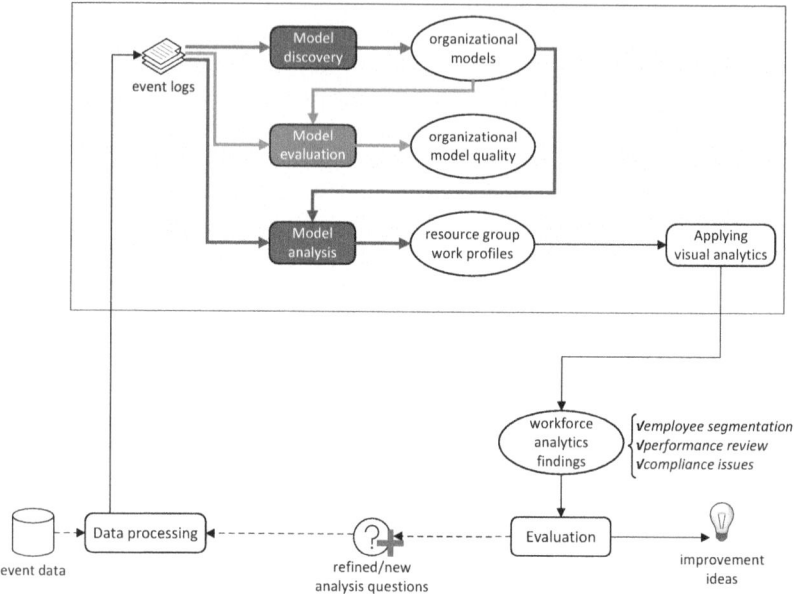

Fig. 1.5: An overview of the proposed approaches illustrated in the context of the Process Mining Project Methodology [38].

How can organizational models be discovered from event log data?

We rigorously define and formalize notions related to event logs and organizational models. For event logs, we assume the existence of some standard data attributes, i.e., case identifier, activity name, timestamp, and resource identifier, which are common for input to organizational model mining. For organizational models, we introduce a new definition that considers multiple process dimensions and, compared to the literature, can more effectively represent the grouping of resources and their involvement in process execution. Then, we use these notions as the foundation for designing and developing our conceptual framework, namely *OrdinoR*[9]. The framework formulates three types of organizational model mining, i.e., model discovery, model evaluation, and model analysis. They all use event logs as input but are performed for different purposes: (i) model discovery aims to construct organizational models to characterize the grouping of resources and their involvement in the actual process execution, reflected by event logs; (ii) model evaluation aims to assess organizational model quality in an objective way, independent from

---

[9] *Ordino* means "to arrange" in Latin; the trailing letter *R* stands for "resources".

1.3 Outlook

additional information other than the input event logs used to discover the models; (iii) model analysis aims to examine the performance of resource groups captured in organizational models and provide information to support workforce analytics. In the framework, we propose that three concrete tasks need to be addressed to discover organizational models. We introduce two measures, fitness and precision, for evaluating organizational models against event logs. We also present a set of quantitative measures for analyzing the behavior of resource groups — their workload distribution and the contribution by group members — based on event logs. Chapter 2 presents the *OrdinoR* framework and these underpinning notions and measures.

Next, we develop an approach to discovering organizational models from event logs. This approach addresses the concrete discovery tasks outlined in the conceptual framework. Specifically, Chapter 3 introduces the problem of learning execution contexts from event logs. Solving this problem is essential to model discovery as well as its evaluation and analysis. We formulate the problem as a task of deriving logical rules that best characterize the specialization of resources recorded in event logs. We then propose two solutions to the problem: (i) a customized decision-tree-based method, which produces locally optimal rules efficiently, and (ii) a simulated-annealing-based method that searches for near-global-optimal rules. Chapter 4 presents the full approach to organizational model discovery, where each of the outlined discovery tasks can be addressed using several alternative methods. These methods can be selected and configured according to the application of the approach, i.e., the process and organization being analyzed, the analytical questions, and available domain knowledge. These methods together constitute a systematic way to automatically construct organizational models that describe resource groupings reflected by event log data.

How can organizational models be exploited to analyze resource group performance in process execution?

Knowing how organizational models may be extracted from event log data and how they represent knowledge about resource groups, the next step is to investigate how these models may be applied to support workforce analytics.

We propose the notion of resource group work profile in Chapter 5. This notion is developed based on reviewing the management literature on human resource performance measurement. It encompasses an array of quantified indicators, which can be extracted from event logs and organizational models and be applied to measure various aspects relevant to how resource groups and their members work in process execution. Furthermore, we also introduce an approach to applying visual analytics to work profiles extracted from data, so as to track, compare, and correlate resource groups' performance — across group and individual levels, over different time periods, and related to various process dimensions.

Together, the conceptual framework and the array of data-driven approaches form a purposeful and viable artifact for organizational model mining in the process mining field.

To demonstrate the usefulness and evaluate the effectiveness of our approaches, we implement them as open-source software tools and conduct experiments on publicly available, real-life event log datasets collected from three different business domains. We exploit the evaluation results to iterate the design of our approaches. The experiments and the results are reported in the evaluation and case study sections of Chapters 3, 4, and 5, respectively.

Finally, we document the steps taken to develop the approaches and share the software implementation as open-source tools.

How may organizational model mining techniques be applied in practice?

As a set of process mining techniques, our approaches can be applied by data analysts following the Process Mining Project Methodology (PM^2) [38]. Refer to Figure 1.5 again.

Here, we assume that the planning and extraction stages have been completed (cf. Section 2 in [38]), so that the event data and process domain knowledge are given with regard to the set of analysis questions and the scope.

In the data processing stage, the application of our approaches requires producing event logs that have at least the minimum set of standard data attributes (which are discussed in detail in Chapter 2). With a created event log, data analysts start by applying the model discovery approach (Chapter 3 and Chapter 4). Then, the quality of the discovered models is determined by applying the model evaluation measures (Chapter 2). The analysts can use the measured quality to decide whether to keep the output models or restart model discovery. In the latter case, they can utilize the model analysis measures (Chapter 2) to "diagnose" the issues of the low-quality models and use the analysis results to adjust the selection and configuration of methods in the next run of the model discovery approach. Once a satisfactory model is obtained, the analysts can apply it along with the event log to extract and analyze resource group work profiles (Chapter 5) using visual analytics.

As outputs, several types of workforce analytics findings can be obtained: (i) the grouping of resources captured by the discovered organizational model may inform decisions about employee segmentation [83]; (ii) the resource group work profiles and their analyses provide an intuitive means to navigate across different aspects, time periods, and multiple process dimensions to objectively review the performance of resource groups and their members in process execution; (iii) the overall examination of the model and the work profiles may reveal potential compliance issues, e.g., an unexpected split of responsibility in handling process activities for specific types of cases, or a failure to ensure agreed workload limits for specific groups of employees.

Lastly, in the evaluation stage, analysts need to validate and interpret the findings and determine if they are useful for supporting improvement ideas; or if it is necessary to perform another iteration of analysis, using questions refined or created during the evaluation.

Chapter 2
Framework for Organizational Model Mining

Abstract In this chapter, we propose *OrdinoR*, a novel framework for organizational model mining. We introduce a new, formalized notion of organizational model as the foundation of the framework. Compared with the state-of-the-art, our notion of organizational model is richer, as it specifies not only resources and their groups but also the connection between resource groups and the multiple dimensions of process execution, captured by the so-called *execution contexts*. We will elaborate on these concepts in Sections 2.2 and 2.3.

Figure 2.1 illustrates the *OrdinoR* framework for organizational model mining. Built upon the new notion of organizational model, the framework is designed to support three types of organizational model mining tasks as follows.

1. *Discovering* organizational models: this task aims to construct models from event logs to reflect the grouping of resources and their involvement in process execution (Section 2.4).
2. *Evaluating* organizational models: this task aims to assess model quality by comparing models against event logs (Section 2.5).
3. *Analyzing* organizational models: this task aims to examine the actual behavior of resource groups captured in organizational models using event logs. Findings from such analyses can be used to provide (i) insights into group-oriented workforce analytics and (ii) diagnostic information explaining the results of evaluating organizational models (Section 2.6).

2.1 Preliminaries

We start by introducing some preliminary concepts and mathematical notation. These preliminaries will be used throughout the remainder of this book.

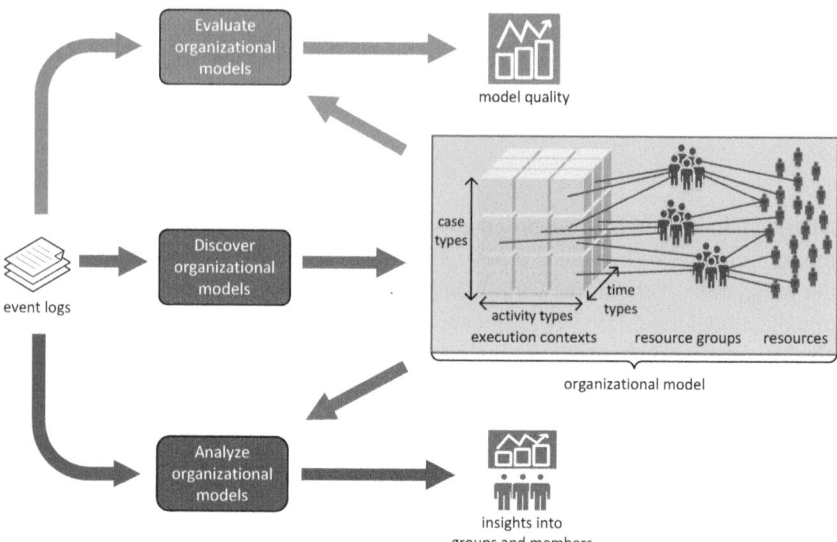

Fig. 2.1: An overview of the *OrdinoR* framework for organizational model mining. It supports three types of mining tasks: discovery, evaluation, and analysis of organizational models using event logs.

Sets and Functions

A *set* is a collection of different elements, which can be objects of any kind, e.g., numbers or symbols. A *universe* is the set of all objects of a certain kind that we wish to consider in a given situation. For example, we may refer to $\mathbb{Z}$, i.e., the set of all integers, as the universe of integers.

We will use conventional set theory notation: $\emptyset$ for the empty set, $\in$ for the containment relation, $\subseteq$ for subset, $\cup$ for set union, $\cap$ for set intersection, and $\setminus$ for set difference. Given a finite set X, $|X|$ denotes the *cardinality* of X, i.e., the number of elements in X. For example, given $X = \{a, b\}$, we have $|X| = 2$. A set is a *singleton* if and only if its cardinality is 1.

$\mathcal{P}(X)$ denotes the *power set* of X, i.e., the set of all subsets of X. $A \in \mathcal{P}(X)$ if and only if $A \subseteq X$. For example, given $X = \{a, b\}$, we have $\mathcal{P}(X) = \{\emptyset, \{a\}, \{b\}, \{a, b\}\}$.

A *partition* of a set is a collection of its non-empty subsets, such that every element in the set is contained in exactly one subset. For the previous example $X = \{a, b\}$, we have $\{\{a\}, \{b\}\}$ and $\{\{a, b\}\}$ as two partitions. Note that the latter can be written as $\{X\}$ — this is called the *trivial partition* of X. For any non-empty set, there exists a trivial partition.

Given two sets X and Y, $X \times Y$ is their *Cartesian product*, i.e., the set of all ordered pairs where the first element is a member of X and the second is a member of Y. For example, with $X = \{2, 3\}$ and $Y = \{p, q\}$, we have $X \times Y = \{ (x, y) \mid x \in X \wedge y \in$

2.1 Preliminaries

Y} = {$(2, p), (2, q), (3, p), (3, q)$}. Cartesian products can be generalized to n sets, and each element of a Cartesian product is therefore a *sequence* of length n, denoted by σ. Specifically, we use $\sigma(i)$ to denote the i-th element of a sequence σ, i.e., σ can be viewed as a function mapping the indexing set of the collection of the n sets onto element values in the sequences. With a slight abuse of notation, let $|\sigma|$ denote the length of σ. For example, for an ordered pair $\sigma = (3, q)$ in the previous example, we have $\sigma(1) = 3$, $\sigma(2) = q$, and $|\sigma| = 2$.

A *multiset* generalizes the concept of a set, allowing multiple occurrences of an element. We denote $\mathcal{B}(X)$ as the set of multisets over X. For example, given $X = \{2, 3\}$, $M = [2^2, 3] \in \mathcal{B}(X)$ is a multiset where element 2 has two occurrences and element 3 has one occurrence. Specifically, we will use *multiset comprehension* $[\, f(x) \mid x \in M \bullet \phi(x)\,]$ to specify multisets where for every $x \in M$ satisfying condition ϕ, an element $f(x)$ is considered to be part of the multiset. For example, $[\, x + 1 \mid x \in [2^2, 3, 5^3] \bullet x < 4 \,]$ yields $[3^2, 4]$.

A *total function* $f: X \rightarrow Y$ maps elements of a set X onto elements of a set Y. X is the *domain* of f, i.e., $dom(f) = X$, and Y is the *codomain* of f. The *range* of f is denoted as $rng(f) = \{\, f(x) \mid x \in X \,\}$, where $f(x)$ is said to be the *image* of x.

f is *injective* if every element in its codomain is the image of at most one element in its domain. f is *surjective* if every element in its codomain is the image of at least one element in its domain. f is *bijective* if it is both injective and surjective — every element in its codomain is the image of exactly one element in its domain (one-to-one correspondence).

A *partial function* $f: X \nrightarrow Y$ maps elements of a subset of X onto elements of Y, i.e., $dom(f) \subseteq X$. $f(x)$ is defined if and only if $x \in dom(f)$.

Let $f: X \rightarrow Y$ be a function and $A \subseteq X$ a subset of X, then the *restriction* of f on A is the function $f\restriction_A: A \rightarrow Y$, with $f\restriction_A(x) = f(x)$ for $x \in A$.

Processes and Human Resources

We will use notions established in the field of business process management and process mining [37, 7]. A process consists of a set of logically connected *activities* performed in an organization and captures possible alternative ways to achieve a business goal. An instance of the execution of a process is a *case*. Process execution involves *resources* performing a sequence of activities in the process. Resources can be individual employees or organizational units such as project teams. Resources can also be machines acting for humans, for example, equipment operated by employees or a workflow system automaton working on behalf of managers.

We will use the term *resource group* to describe entities that represent the grouping of resources. Note that a resource group may refer to either a group of individuals, e.g., a business role or position held by employees, or a group of organizational units, e.g., a large department that contains several smaller units. The interpretation of resource groups depends on the interpretation of the resources being analyzed.

Event Logs

Data related to process execution is recorded by process-aware information systems [7], notably in the form of *event logs*. An event log consists of a set of timestamped events with a range of *event attributes*, providing factual information on how activities were carried out by resources participating in process execution.

Table 2.1 shows an example event log that records the execution of an insurance claim handling process. Rows correspond to events and columns correspond to event attributes. For instance, the first row is an event recording that a resource named "Pete" registered a request for an insurance claim with id "654422", which is related to a "bronze" customer. This activity was not performed on-site, and the timestamp "29-08-2018 13:36" records the time when it happened.

Table 2.1: A fragment of an example event log.

case id	activity name	timestamp	resource	customer type	on-site
...	...	...	...	...	...
654422	register request	29-08-2018 13:36	Pete	bronze	no
654423	register request	29-08-2018 15:02	Pete	silver	no
654424	register request	29-08-2018 16:08	Pete	silver	no
654422	confirm request	29-08-2018 16:10		bronze	
654424	confirm request	29-08-2018 16:12		silver	
654423	get missing info	29-08-2018 16:28	Ann	silver	no
654423	confirm request	29-08-2018 16:45		silver	
654423	check insurance	30-08-2018 09:09	John	silver	no
654424	check insurance	30-08-2018 09:22	Sue	silver	yes
654425	register request	30-08-2018 10:07	Bob	gold	no
654423	accept claim	30-08-2018 11:32	John	silver	no
654424	reject claim	30-08-2018 11:45	Sue	silver	no
654423	pay claim	30-08-2018 11:48		silver	
654425	confirm request	30-08-2018 12:44		gold	
654425	check insurance	30-08-2018 13:32	Mary	gold	yes
654425	accept claim	30-08-2018 14:09	Mary	gold	no
654425	pay claim	30-08-2018 14:14		gold	
...	...	...	...	...	...

We define a general data structure for event logs (Definition 2.1). An event log (EL) contains a set of uniquely identifiable events (E), a set of event attribute names (Att), and the corresponding event attribute values carried by each event (as specified by function π). It is possible that an event does not carry any value for a given event attribute, e.g., in Table 2.1 there are events with no resource information. Hence, function π is a partial function mapping the attributes of events to values.

Definition 2.1 (Event Log) $\mathcal{E}$ is the universe of event identifiers, $\mathcal{U}_{Att}$ is the universe of possible attribute names, and $\mathcal{U}_{Val}$ is the universe of possible attribute values. An event log is a tuple $EL = (E, Att, \pi)$ with $E \subseteq \mathcal{E}$, $E \neq \emptyset$, $Att \subseteq \mathcal{U}_{Att}$, and

$\pi : E \to (Att \nrightarrow \mathcal{U}_{Val})$. An event $e \in E$ has attributes $dom(\pi(e))$. For an attribute $x \in dom(\pi(e))$, $\pi_x(e) = \pi(e)(x)$ is the attribute value of x for event e.

Next, we elaborate on the definition of event attributes needed for storing the essential information about process execution (Definition 2.2). An event log usually records multiple cases. Each case can be uniquely identified and is related to a sequence of events corresponding to activities executed at some specific time. As the minimum requirement for event logs, events have three standard attributes: case identifier (*case*), activity name (*act*), and timestamp (*time*). Optionally, an event records the resource (*res*) who performed the activity. In addition to these four common attributes, an event log may record event attributes such as *customer type* and *cost*, which vary across different processes and information systems. Note that an event attribute is considered a discrete attribute if it has a finite or countably infinite set of values, e.g., *customer type* may record only a limited set of pre-defined customer type names ("bronze", "silver", and "gold" in the example log); otherwise, it is a continuous attribute, e.g., *cost* may record real numbers that are valid in the context of the corresponding business process.

Specifically, we will say that an event attribute is a case attribute if events belonging to the same case all share an identical value for that attribute. For example, case identifier is a case attribute.

Definition 2.2 (Event Attributes) Let $C \subseteq \mathcal{U}_{Val}$, $\mathcal{A} \subseteq \mathcal{U}_{Val}$, $\mathcal{T} \subseteq \mathcal{U}_{Val}$, and $\mathcal{R} \subseteq \mathcal{U}_{Val}$ denote the universes of case identifiers, activity names, timestamps, and resource identifiers, respectively. Any event log $EL = (E, Att, \pi)$ has three special attributes from the set $D = \{case, act, time\}$, referred to as the core event attributes, and a special attribute *res*, i.e., $D \cup \{res\} \subseteq Att$, such that for any $e \in E$:

- $D \subseteq dom(\pi(e))$,
- $\pi_{case}(e) \in C$ is the case to which e belongs,
- $\pi_{act}(e) \in \mathcal{A}$ is the activity e refers to,
- $\pi_{time}(e) \in \mathcal{T}$ is the time at which e occurred, and
- $\pi_{res}(e) \in \mathcal{R}$ is the resource that executed e if $res \in dom(\pi(e))$.

Given a resource $r \in \mathcal{R}$, let $[E]_r = \{e \in E \mid res \in dom(\pi(e)) \wedge \pi_{res}(e) = r\}$ denote the set of events in the log executed by that resource. $[E]_\mathcal{R} = \bigcup_{r \in \mathcal{R}} [E]_r$ is the set of all events in the log that have resource information.

2.2 Execution Context

A key feature of the organizational models proposed in our research is the ability to capture the involvement of resource groups in process execution. This is achieved through the notion of *execution context*.

In business process execution, the groupings of resources are often associated with certain contexts, as reflected in event logs by the different types of activities or cases performed by resources, or the times when resources perform activities [77].

Consider the example event log of an insurance claim handling process in Table 2.1. Pete and Bob only performed activity "register (a claim) request", while John, Sue, and Mary performed "check insurance" and decided whether to "accept" or "reject" a claim. This may be related to the different business roles of these employees. In the meantime, Bob and Mary only handled a claim from a "gold" customer, while others only handled claims from "silver" customers. This can be a result of the insurance company setting up separate teams serving "gold" customers.

To reach such findings in the example above, it is essential to categorize events into *types* and use them to capture those different contexts. We consider *case types*, *activity types*, and *time types* (Definition 2.3) related to the three core dimensions of process execution. This is inspired by the research on process cubes [6], which proposes to organize events by different dimensions to enable a systematic view and analysis of large-scale, multidimensional event data. Case types describe the categories of cases, for example, insurance claims can be classified by the type of customers who lodged the claims (e.g., considering "gold" customers as VIPs and other types as normal customers). Similarly, activity types categorize activity names into groups of relevant activities (e.g., registration, approval) and time types categorize timestamps into periods (e.g., weekdays vs. weekends, morning vs. afternoon).

Definition 2.3 (Case Types, Activity Types, and Time Types) Let CT, AT, and TT denote the sets of names of case types, activity types, and time types, respectively. The functions $\varphi_{case}: CT \to \mathcal{P}(C)$, $\varphi_{act}: AT \to \mathcal{P}(A)$, and $\varphi_{time}: TT \to \mathcal{P}(T)$ define partitions over C, A, and T, respectively. We will use a special type $\bot$ that is associated with all cases, all activities, and all times, i.e., $\varphi_{case}(\bot) = C$, $\varphi_{act}(\bot) = A$, and $\varphi_{time}(\bot) = T$. The sets CT, AT, and TT only share this special type and are otherwise mutually disjoint. We define $CO = CT \times AT \times TT$.

We now formalize the notion of *execution context* (Definition 2.4) as consisting of a case type, an activity type, and a time type. Each execution context characterizes a possible way of executing an activity in a process and can be associated with a specific set of events that share similar characteristics. For instance, we can relate the first two events in the example log (Table 2.1) to the same execution context (normal case, registration activity, Wednesday). Note that an execution context can be specified with a "wild card" for any of its constituent components of case type, activity type, and time type. The $\bot$ symbol (formally, as per Definition 2.3, $\bot$ is a case type, an activity type, and a time type) is used for those components that are not meant to be restricting. Consider for example the execution context (normal case, $\bot$, Wednesday). This execution context concerns all process activities that are executed on Wednesdays when handling insurance claims from normal customers, i.e., those of "silver" or "bronze" type, following our previous example.

Definition 2.4 (Execution Context) An execution context co is an element of CO. Given an event log $EL = (E, Att, \pi)$ and an execution context $co = (ct, at, tt)$,

$$[E]_{co} = \{ e \in E \mid \pi_{case}(e) \in \varphi_{case}(ct) \land \pi_{act}(e) \in \varphi_{act}(at) \land \pi_{time}(e) \in \varphi_{time}(tt) \}$$

is the set of events in the log having that execution context.

2.3 Organizational Model

Figure 2.2 illustrates the notion of execution context. In Figure 2.2a, events are seen as data points in a three-dimensional space capturing information on cases, activities, and time. An event may be related to an individual resource executing an activity. In Figure 2.2b, event attribute values on each of the three dimensions are partitioned by some specified collection of case types, activity types, and time types. Each combination of a case type, an activity type, and a time type specifies an execution context, represented as a "cube" in the data space. Resources that originated events from the same cubes may belong to the same resource group.

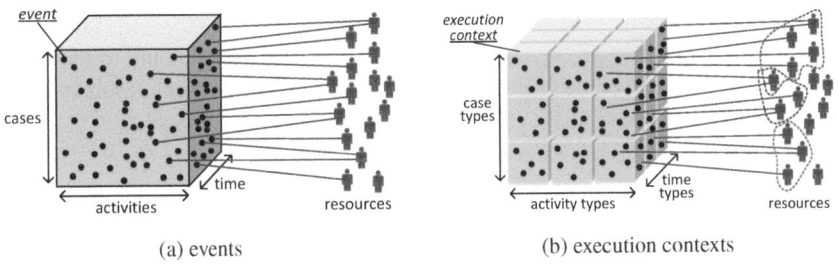

(a) events   (b) execution contexts

Fig. 2.2: Illustration of (a) events as data points in three-dimensional space along the dimensions of case, activity, and time, and (b) execution contexts as "cubes" characterized by case types, activity types, and time types.

2.3 Organizational Model

Our notion of *organizational model* (Definition 2.5) incorporates the concept of execution context. While this model contains, as per usual, the resource groups (*RG*) and their members (*mem*), it further captures the involvement of resource groups in process execution, i.e., the "capabilities" of groups, by linking groups with execution contexts (*cap*).

Definition 2.5 (Organizational Model) Let $\mathcal{R}$ be the universe of resource identifiers. An organizational model is a tuple $OM = (RG, mem, cap)$ where RG is a set of resource groups, $mem: RG \to \mathcal{P}(\mathcal{R})$ maps each resource group onto its members, and $cap: RG \to \mathcal{P}(CO)$ maps each resource group onto its possible execution contexts.

Figure 2.3 illustrates the proposed notion of organizational model. The many-to-many relationships capture the fact that a resource may belong to multiple groups and a resource group may be associated with multiple execution contexts. Formally, there may exist two distinct groups rg_1 and rg_2 such that $mem(rg_1) \cap mem(rg_2) \neq \emptyset$ and $cap(rg_1) \cap cap(rg_2) \neq \emptyset$.

24   2 Framework for Organizational Model Mining

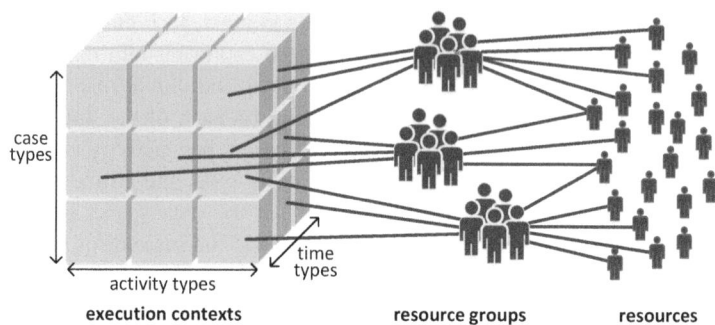

Fig. 2.3: Illustration of an organizational model which captures many-to-many relationships between resource groups and resources and those between resource groups and execution contexts.

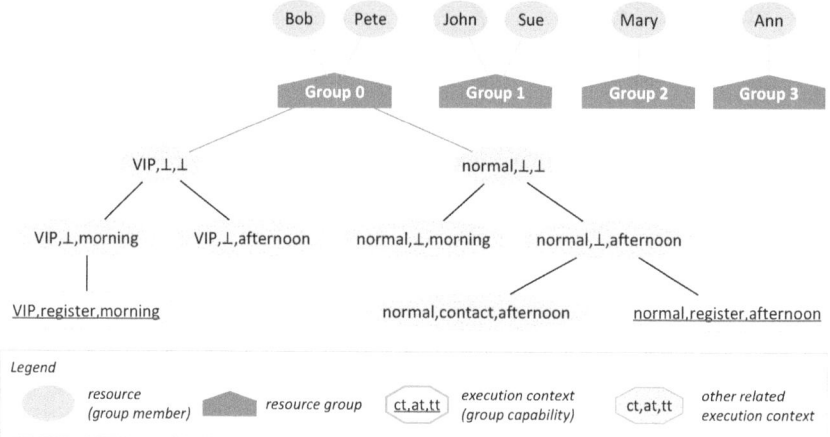

Fig. 2.4: Visualization of an example organizational model related to the event log in Table 2.1.

Figure 2.4 depicts the visualization of an example organizational model, in which different colored shapes represent resources, resource groups, and their related execution contexts, respectively. For instance, "Group 0" has two member resources, Bob and Pete, who are capable of executing activities related to execution contexts "(VIP, register, morning)" and "(normal, register, afternoon)". These two execution contexts, as the group's capabilities, are highlighted in red in the visualization, with the type names underlined.

2.4 Discovering Organizational Models

To discover organizational models from an event log, it is useful to view events as samples of resource behavior in process execution [62]. We use the term *resource event* to denote an event that captures a resource's involvement in some execution context. A *resource-event log* (Definition 2.6) is a multiset of resource events and represents a resource view on process execution data through execution contexts, i.e., a resource-event log records how some resources performed certain activities for certain cases at certain times when they participated in the execution of a process. A resource-event log can be derived from an event log using a collection of execution contexts (Definition 2.7).

Definition 2.6 (Resource-Event Log) A resource event is a tuple $(r, co) \in \mathcal{R} \times CO$. A resource-event log $RL \in \mathcal{B}(\mathcal{R} \times CO)$ is a multiset of resource events.

Definition 2.7 (Derived Resource-Event Log) Let $EL = (E, Att, \pi)$ be an event log and let $CO \subseteq \mathcal{CO}$ be a pre-defined collection of execution contexts. The resource-event log derived from EL and CO is $RL(EL, CO) = [\,(\pi_{res}(e), co) \mid co \in CO, e \in [E]_{\mathcal{R}} \bullet e \in [E]_{co}\,]$.

Table 2.2 shows an example resource-event log derived from the event log in Table 2.1, using execution contexts defined based on the case types, activity types, and time types below.

- Two case types are defined based on the event attribute *customer type*, which distinguishes two groups of customers, namely normal (for "silver" and "bronze" customers) and VIP (for "gold" customers). Therefore, {"654422", "654423", "654424"} $\subseteq \varphi_{case}$(normal) and {"654425"} $\subseteq \varphi_{case}$(VIP).
- Four activity types are defined: register, contact, check, and decide. Therefore, {"register request", "confirm request"} $\subseteq \varphi_{act}$(register), {"get missing info", "pay claim"} $\subseteq \varphi_{act}$(contact), {"check insurance"} $\subseteq \varphi_{act}$(check), {"accept claim", "reject claim"} $\subseteq \varphi_{act}$(decide).
- Two time types are defined by dividing working hours in a day into two time frames, namely morning and afternoon. Therefore, timestamps of events are categorized accordingly, e.g., "30-08-2018 09:09" $\in \varphi_{time}$(morning), "29-08-2018 15:02" $\in \varphi_{time}$(afternoon).

Note that a resource event can have multiple occurrences. For example, the first three rows in Table 2.2 both refer to the same resource event "(Pete, normal, register, afternoon)", indicating that Pete conducted an activity in the same execution context three times.

Organizational models can be discovered from an event log based on the similarities of resources characterized by a corresponding derived resource-event log. To do this, the following tasks need to be addressed:

1. *Determine execution contexts* by specifying the relevant case types, activity types, and time types based on the input event log;

Table 2.2: A fragment of an example derived resource-event log.

resource	case type	activity type	time type
...	...	...	...
Pete	normal	register	afternoon
Pete	normal	register	afternoon
Pete	normal	register	afternoon
Ann	normal	contact	afternoon
John	normal	check	morning
Sue	normal	check	morning
Bob	VIP	register	morning
John	normal	decide	morning
Sue	normal	decide	morning
Mary	VIP	check	afternoon
Mary	VIP	decide	afternoon
...	...	...	...

2. *Determine resource grouping* by identifying clusters of resources who share similar behavior in process execution;
3. *Determine the links between resource groups and execution contexts* to describe the involvement of resource groups in process execution.

In Chapters 3 and 4, we will develop approaches for addressing these tasks, respectively.

2.5 Evaluating Organizational Models

As discussed in the literature review, it remains an open issue how to evaluate discovered organizational models against input event logs. We address this gap by introducing two notions to organizational model mining, namely *fitness* and *precision* (cf. [7]), and their corresponding quantitative measures. The two notions are based on the new definition of organizational model and provide two perspectives for assessing an organizational model with respect to an event log.

2.5.1 Fitness

Fitness evaluates the completeness [42] of a model with respect to a log, i.e., to what degree behavior observed in the log is allowed by the model. To quantify fitness, we first introduce the notion of *conforming events* (Definition 2.8). Given a log and a model, an event in the log is conforming if its originating resource is allowed by the model to execute it.

2.5 Evaluating Organizational Models

We define a measure for fitness (Definition 2.9), which yields a value between 0 and 1. Note that only events with resource information (i.e., events in $[E]_\mathcal{R}$) should be considered (hence fitness is only defined if there are events with resource information in the event log).

Definition 2.8 (Conforming Events) Let $EL = (E, Att, \pi)$ be an event log and let $OM = (RG, mem, cap)$ be an organizational model.

$$E_{conf} = \left\{ e \in [E]_\mathcal{R} \mid \exists_{rg \in RG, co \in cap(rg)} [\pi_{res}(e) \in mem(rg) \wedge e \in [E]_{co}] \right\}$$

is the set of all conforming events. $E_{nconf} = [E]_\mathcal{R} \setminus E_{conf}$ consists of all non-conforming events.

Definition 2.9 (Fitness) Let $EL = (E, Att, \pi)$ be an event log with $[E]_\mathcal{R} \neq \emptyset$. The fitness of an organizational model OM with respect to event log EL is

$$fitness(EL, OM) = \frac{|E_{conf}|}{|[E]_\mathcal{R}|}.$$

The fitness between a model and a log is good when most events in the log are conforming events. $fitness(EL, OM) = 1$ if resources only performed events in EL that they were allowed to perform according to OM. $fitness(EL, OM) = 0$ if no event in EL was executed by a resource actually allowed to perform it according to OM. Following the definitions, all events with resource information in the example event log (Table 2.1) are conforming events. Hence, the example organizational model shown in Figure 2.4 has a fitness of 1 with respect to the example event log.

2.5.2 Precision

Precision evaluates the exactness [42] of a model with respect to a log, i.e., the extent to which behavior allowed by the model is observed in the log. To quantify precision, we propose the notion of *candidate resources* (Definition 2.10). Given a log and a model, the candidate resources of an event refer to resources in the model who are allowed to perform the event. The idea is that a perfectly precise model allows exactly the behavior described in the log.

Definition 2.10 (Candidate Resources) Let $EL = (E, Att, \pi)$ be an event log and let $OM = (RG, mem, cap)$ be an organizational model. $cand: E \to \mathcal{P}(\mathcal{R})$ maps events onto sets of candidate resources (possibly empty). For each $e \in E$,

$$cand(e) = \left\{ r \in \mathcal{R} \mid \exists_{rg \in RG, co \in cap(rg)} [r \in mem(rg) \wedge e \in [E]_{co}] \right\}$$

is the set of candidate resources for event e. $cand(E) = \bigcup_{e \in E} cand(e)$ is the overall set of candidate resources.

We also introduce the notion of *allowed events* (Definition 2.11). Given a log and a model, an event in the log is an allowed event if it has at least one candidate resource in the model. Note that if a candidate resource of an event is the originating resource of the event, then such an event is both a conforming event and an allowed event.

Definition 2.11 (Allowed Events) Let $EL = (E, Att, \pi)$ be an event log and let $OM = (RG, mem, cap)$ be an organizational model. $E_{allowed} = \{\, e \in [E]_R \mid cand(e) \neq \emptyset \,\}$ is the set containing all allowed events.

Based on the above, the precision of a model with respect to an event log can be measured by considering the fraction of resources allowed by the model to perform events in the log (Definition 2.12). Like the fitness measure, the precision measure also yields a value between 0 and 1. Note that precision is only defined if there are allowed events in an event log according to a model.

Definition 2.12 (Precision) Let $EL = (E, Att, \pi)$ be an event log and let $OM = (RG, mem, cap)$ be an organizational model, with $E_{allowed} \neq \emptyset$. The precision of organizational model OM with respect to event log EL is

$$precision(EL, OM) = \frac{1}{|E_{allowed}|} \sum_{e \in E_{conf}} \frac{|cand(E)| - |cand(e)| + 1}{|cand(E)|}.$$

Accordingly, $precision(EL, OM) = 1$ if and only if every allowed event in EL is a conforming event and each of them has no other candidate resource than the one who executed the event. On the other hand, $precision(EL, OM) = 0$ if and only if none of the allowed events is a conforming event. For instance, given the organizational model in Figure 2.4, the first event in the example log in Table 2.1 "(654422, register request, 29-08-2018 13:36, Pete, bronze)" has two candidate resources, Bob and Pete, and all events with resource information are allowed events. The precision of this model with respect to the log is 0.879, suggesting that the model allows extra behavior to happen, in addition to that recorded in the log.

For an organizational model discovered from an event log, fitness and precision can be used to assess its quality in terms of how it captures the information recorded in the log, i.e., the reality. A good discovered model is expected to describe the reality both completely (achieving high fitness) and exactly (achieving high precision). Fitness and precision can be incorporated into a single measure for an overall evaluation, e.g., by calculating the F1-score [42].

2.6 Analyzing Organizational Models

In this section, we discuss how organizational models can be analyzed to examine the behavior of resource groups. An organizational model outlines the grouping of resources and their capabilities in terms of process execution. We can extend a model by using event frequencies and temporal information about cases in an event log, and thus "replay" how resource groups in the model and their members participated

2.6 Analyzing Organizational Models

in a process. As a starting point, we introduce four quantitative measures that can be used for analyzing an organizational model. Note that all these measures only apply to events with resource information in an event log, i.e., events in $[E]_\mathcal{R}$.

Group relative focus (Definition 2.13) specifies how much of the overall work by a resource group was performed in an execution context. It can be used to measure how a resource group distributed its workload across different execution contexts, i.e., work diversification of a group. Note that group relative focus is only defined if there are events executed by some member of the group.

Definition 2.13 (Group Relative Focus) Given event log $EL = (E, Att, \pi)$, execution contexts CO, and organizational model $OM = (RG, mem, cap)$, for any resource group $rg \in RG$ with $\{e \in [E]_\mathcal{R} \mid \pi_{res}(e) \in mem(rg)\} \neq \emptyset$, its relative focus on execution context $co \in CO$ can be measured by

$$RelFocus(rg, co) = \frac{|\{e \in [E]_\mathcal{R} \mid \pi_{res}(e) \in mem(rg) \wedge e \in [E]_{co}\}|}{|\{e \in [E]_\mathcal{R} \mid \pi_{res}(e) \in mem(rg)\}|}.$$

Group relative stake (Definition 2.14) specifies how much of the work performed in an execution context was done by a resource group. It can be used to measure how the workload devoted to an execution context was distributed across different resource groups in an organizational model, i.e., work participation by the groups. Note that group relative stake is only defined if there are events having the execution context.

Definition 2.14 (Group Relative Stake) Given event log $EL = (E, Att, \pi)$, execution contexts CO, and organizational model $OM = (RG, mem, cap)$, for any resource group $rg \in RG$, its relative stake in execution context $co \in CO$, with $[E]_\mathcal{R} \cap [E]_{co} \neq \emptyset$, can be measured by

$$RelStake(rg, co) = \frac{|\{e \in [E]_\mathcal{R} \mid \pi_{res}(e) \in mem(rg) \wedge e \in [E]_{co}\}|}{|\{e \in [E]_\mathcal{R} \mid e \in [E]_{co}\}|}.$$

Group coverage (Definition 2.15) specifies the proportion of members of a resource group that performed in an execution context. Note that group coverage is only defined if there are resources in the resource group.

Definition 2.15 (Group Coverage) Given event log $EL = (E, Att, \pi)$, execution contexts CO, and organizational model $OM = (RG, mem, cap)$, for any resource group $rg \in RG$ with $mem(rg) \neq \emptyset$, the proportion of group members that performed in an execution context $co \in CO$ can be measured by

$$Cov(rg, co) = \frac{|\{r \in mem(rg) \mid \exists_{e \in [E]_\mathcal{R} \cap [E]_{co}} \pi_{res}(e) = r\}|}{|mem(rg)|}.$$

Group member contribution (Definition 2.16) specifies how much work conducted in an execution context by a group was performed by a specific group member. It can be used to measure how a group's workload related to an execution context was distributed across its members. Note that group member contribution is only defined if at least one member of the resource group executed an event having the execution context.

Definition 2.16 (Group Member Contribution) Given event log $EL = (E, Att, \pi)$, execution contexts CO, and organizational model $OM = (RG, mem, cap)$, for resource group $rg \in RG$ and execution context $co \in CO$, with $\{ e \in [E]_\mathcal{R} \cap [E]_{co} \mid \pi_{res}(e) \in mem(rg) \} \neq \varnothing$, the contribution of a group member $r \in mem(rg)$ can be measured by

$$MemContr(r, rg, co) = \frac{|\{ e \in [E]_\mathcal{R} \mid \pi_{res}(e) = r \land e \in [E]_{co} \}|}{|\{ e \in [E]_\mathcal{R} \mid \pi_{res}(e) \in mem(rg) \land e \in [E]_{co} \}|}.$$

Consider the example organizational model in Figure 2.4. For resource group "Group 0" and one of its capabilities (VIP, register, morning), we have:

- *RelFocus*("Group 0", (VIP, register, morning)) = 0.25,
- *RelStake*("Group 0", (VIP, register, morning)) = 1.0,
- *Cov*("Group 0", (VIP, register, morning)) = 0.5,
- *MemContr*(Bob, "Group 0", (VIP, register, morning)) = 0,
- *MemContr*(Pete, "Group 0", (VIP, register, morning)) = 1.0.

As shown above, resources in "Group 0" devoted 25% of their total workload (indicated by *RelFocus*) to carrying out activities related to "registering requests for VIP cases in the morning"; "Group 0" is the only group that contributed to such work (indicated by *RelStake*) in the process; only 50% of the group members (indicated by *Cov*) actually participated in this type of work, and that is resource Pete (indicated by *MemContr*).

In addition to providing insights into the performance of resource groups and their members in process execution, model analysis measures can also be used for "diagnosing" the differences between an organizational model and a log. In the example organizational model, we can observe that "Group 0" is the only one that has a comparatively low group coverage in terms of its capabilities — the model considers *both* of its members, Bob and Pete, capable of performing in *both* execution contexts, but the event log does not show such behavior. This explains the imperfect precision (0.879) of the model. Also, if the example model is a discovered model, then the revealed differences may inform how the discovery method could be improved to construct more precise models.

2.7 Discussion

This chapter presents a conceptual framework, *OrdinoR*, for discovering, evaluating, and analyzing organizational models using event logs. The framework is built around a rich notion of organizational model, where resource groups are linked to execution contexts that capture employees' capabilities of performing different types of activities or cases at different time periods. Based on that, organizational models

2.7 Discussion

can be discovered from event logs to reflect the grouping of resources and their involvement in process execution. The *OrdinoR* framework also introduces fitness and precision. The two measures provide a rigorous means for evaluating the quality of an organizational model against an event log, based on the extent to how completely and exactly the model can describe the log observations. Last but not least, the framework presents measures for model analysis, which allows an organizational model to be extended with log data to examine the actual behavior of resource groups captured by the model.

Our proposal of the *OrdinoR* organizational model mining framework contributes a novel idea to the research area. First, compared to models in the literature, the new notion of organizational model characterizes the capabilities of resource groups via execution contexts. This feature allows discovered organizational models to be used for explaining the grouping of resources and linking the grouping to process execution. Therefore, the *OrdinoR* models are more effective in terms of enabling analysis of resource group performance. Moreover, the proposal of *OrdinoR* introduces novel tasks and challenges regarding how these organizational models may be discovered from event logs, as will be shown in the following chapters. Second, the model evaluation measures compare discovered models with the event logs used as input for discovery, and independently of *how* the models are discovered. As such, *OrdinoR* enables *extrinsic* evaluation of discovered organizational models.

There are many possibilities to advance the *OrdinoR* framework. One such possibility would be to extend the notion of execution context. The current definition considers only the three core dimensions of process execution, i.e., case, activity, and time, which are in line with the minimum requirement for event logs. When additional event attributes are recorded in event logs, it is possible to consider other dimensions that are useful for characterizing the execution of process activities — for example, location of resources originating activities or costs required to complete activities. Execution contexts with more dimensions can enable a richer modeling of resource groups and their capabilities.

Chapter 3
Learning Execution Contexts

Abstract Execution contexts form a key component in the *OrdinoR* organizational models. This notion enables us to systematically combine multidimensional information in event logs and use that to precisely characterize the involvement of resource groups in process execution. In Chapter 2, we showed examples of case types, activity types, and time types, and how they can be combined to define execution contexts. Those examples can be seen as a result of *manually* specifying execution contexts based on prior information, such as domain knowledge about an event log and given analysis questions. In this chapter, we introduce an approach that supports *automatically learning* execution contexts from an event log and explain why it is desirable to have such an approach.

Let us revisit the example in Section 2.4. The four activity types imply the existence of an abstract view of the insurance claim process, e.g., "accept claim" and "reject claim" are grouped by type "decide" as they are variants of decisions made on insurance claims, "get missing info" and "pay claim" are grouped by "contact" as both are likely to involve contacting the customer. This abstraction is not directly recorded in the event log (Table 2.1) but may be understood by process owners or analysts who possess relevant domain knowledge. The two time types correspond to a selected level of granularity of timestamps. This categorization of events may be guided by some questions focused on analyzing the performance of human resources during different working hours (morning vs. afternoon). While domain knowledge and guiding questions are key to analyses of event logs, they cannot be assumed to be readily available or sufficiently concrete [80, 38]. Therefore, manually defining execution contexts — as shown in the example — may not always be an option.

In the following sections, we will introduce a learning approach that aims at exploiting the discriminative information of events embedded in the data rather than relying on prior information. The approach requires minimal user input and is capable of automatically extracting a set of logic rules from an event log, which can then be used to define high-quality execution contexts.

© The Author(s), under exclusive license to Springer Nature Switzerland AG 2026
R. J. Yang, *Mastering Organizational Dynamics Using Process Mining*, LNBIP 552, pp. 32–60, 2026.
https://doi.org/10.1007/978-3-031-93530-5_3

3.1 Preliminaries

Our approach utilizes the discriminative information of events concerning resources — that is, patterns showing event attribute values that exclusively or frequently coappear and in combination with certain resource identifiers. Recall the previous observation on the example event log (Table 2.1): (i) activity "register request" only appeared with resources Pete and Bob; (ii) activity "check insurance" only appeared with resource Mary when the customer type is "gold" and otherwise with John and Sue. Patterns like these are often the results of the division of labor among resources working in process execution, and our approach is designed to utilize such information to construct execution contexts.

However, given an event log, not all event attributes can be exploited. Only those satisfying the requirements of *type-defining attributes* (Definition 3.1) may be used to define a set of types for one of the process execution dimensions (i.e., case, activity, and time).

Definition 3.1 (Type-Defining Attributes) Given an event log $EL = (E, Att, \pi)$ with $D = \{case, act, time\}$ the core event attributes. For any $e \in E$ and $d \in D$, let

- $X \subseteq dom(\pi(e))$ be some event attributes recorded in the log,
- $\pi(e)\restriction_X$ the restriction of $\pi(e)$ on X, and
- $V = \{\pi(e)\restriction_X \mid e \in E\}$ the mappings of the attributes in X recorded in EL.

Then X is a set of type-defining attributes for d in EL, if and only if there exists a non-trivial partition P of V, such that for all $p, q \in P$,

$$p \neq q \Rightarrow \{\pi_d(e) \mid e \in E \land \pi(e)\restriction_X \in p\} \cap \{\pi_d(e) \mid e \in E \land \pi(e)\restriction_X \in q\} = \emptyset,$$

i.e., the partition P corresponds to a partition of the set of distinct values of d recorded in EL.

In the example event log (Table 2.1), case attribute *customer type* is a type-defining attribute for case types — each case records at most one customer type value, and any partitioning over customer types will result in distinct groups of cases. Note that multiple type-defining attributes can be used for a process dimension. For instance, a case attribute *insurance type* may exist in the example log and the values can be combined with *customer type* values to categorize cases into disjoint groups, e.g., ("gold", "health insurance") and ("silver", "car insurance"/"boat insurance"). By contrast, consider another example: for activity types, event attribute *on-site* alone does not qualify as a type-defining attribute — activity name "check insurance" may correspond to either "yes" or "no" for *on-site*. Hence, there exists no partition over the attribute values of *on-site* that can induce a partition of the set of distinct activity names.

By this definition, any core event attribute can be used as a type-defining attribute for itself.

3.2 Problem Modeling

In this section, we introduce how we model the problem of learning execution contexts from an event log. We first present the idea of *categorization rules* that represent the classification of case types, activity types, and time types. Then, we discuss how to measure the *quality* of execution contexts with regard to an event log. Finally, we formulate the execution context learning problem based on the notion of categorization rules and the proposed quality measures.

3.2.1 Categorization Rules

A set of execution contexts specifies a way of partitioning events by defining case types, activity types, and time types. Hence, learning execution contexts from an event log requires learning those types, i.e., the classification of cases, activities, and times. To this end, we propose to use *categorization rules* to represent types.

A categorization rule (Definition 3.2) is a Boolean formula in conjunctive normal form, consisting of one or more clauses. Each clause can evaluate an event by its value of some event attribute. For instance, *customer type* $\in$ {normal} $\wedge$ *cost* $\in$ $[10000, \infty)$ is a categorization rule that evaluates a discrete attribute *customer type* and a continuous attribute *cost*. Given a set of events, evaluating this rule filters events that record normal customers and cost greater than or equal to 10000.

Definition 3.2 (Categorization Rule) Given an event log $EL = (E, Att, \pi)$, let $d \in D$ be a core event attribute, and let $X \subseteq Att$ be a set of type-defining attributes for d. $\phi = \bigwedge_{x \in X} \bar{x} \lessdot \bar{U}_x$ is a categorization rule, where $U_x \in \mathcal{P}(\mathcal{U}_{Val})$ is a set of attribute values for an attribute $x \in X$. For any $e \in E$, ϕ can be evaluated as follows: $[\![\phi]\!](e) = \bigwedge_{x \in X} [\![\bar{x} \lessdot \bar{U}_x]\!](e) = \bigwedge_{x \in X} \pi_x(e) \in U_x$.

- $[E]_\phi = \{e \in E \mid [\![\phi]\!](e)\}$ is the set of events in the log satisfying the categorization rule ϕ.
- We introduce a default rule true such that $[\![\text{true}]\!](e) = true$ for all $e \in E$. It follows that $[E]_{\text{true}} = E$.
- Any two categorization rules ϕ_1 and ϕ_2 are equivalent, i.e., $\phi_1 \cong \phi_2$, if and only if $[E]_{\phi_1} = [E]_{\phi_2}$ for any $E \subseteq \mathcal{E}$. Otherwise, we write $\phi_1 \not\cong \phi_2$.

A set of categorization rules can be used to define a set of types on an event log (Definition 3.3). Consider the example of grouping customer types into normal customers and VIPs to define case types. This can be defined as a set of rules $\Phi_1 = \{customer\ type \in \{silver, bronze\}, customer\ type \in \{gold\}\}$, as long as a customer can only be either gold, silver, or bronze. But, another example of three rules, $\Phi_2 = \{customer\ type \in \{gold\}, customer\ type \in \{silver\}, customer\ type \in \{bronze\}\}$, would also define different case types that are specific to each customer type.

3.2 Problem Modeling

Definition 3.3 (Defining Types as Categorization Rules) Given an event log $EL = (E, Att, \pi)$, let $d \in D$ be a core event attribute. Φ is a set of categorization rules that define a set of types on d, if and only if:

1. for any $\phi_1, \phi_2 \in \Phi$ with $\phi_1 \neq \phi_2$, $\{\pi_d(e) \mid e \in [E]_{\phi_1}\} \cap \{\pi_d(e) \mid e \in [E]_{\phi_2}\} = \emptyset$; and
2. $\bigcup_{\phi \in \Phi} \{\pi_d(e) \mid e \in [E]_\phi\} = \bigcup_{e \in E} \{\pi_d(e)\}$,

i.e., the subsets of events satisfying categorization rules in Φ induce a partition of all values of d recorded in EL.

Execution contexts can be defined by three sets of categorization rules that define case types, activity types, and time types, respectively (Definition 3.4). Note that the use of categorization rules provides a different representation from the one we introduced. In Chapter 2, we formalized the concept of execution contexts based on the names of types and their mapping to the core event attributes (*case, act, time*). That representation serves as a general data structure for execution contexts, but does not associate them with other event attributes (e.g., *customer type*) in a given event log. Here, the use of categorization rules enables a clear connection between the input (event attributes and values in a log) and the output (execution contexts), which is essential to solving the learning problem.

Definition 3.4 (Defining Execution Contexts as Categorization Rules) Given an event log $EL = (E, Att, \pi)$, let Φ_{case}, Φ_{act}, and Φ_{time} be three sets of categorization rules that define case types, activity types, and time types, respectively. $CO = \Phi_{case} \times \Phi_{act} \times \Phi_{time}$ is a set of execution contexts defined by the three sets of categorization rules.

CO specifies a way of partitioning EL. Given an execution context $co = (\phi_c, \phi_a, \phi_t) \in CO$, $[E]_{co} = [E]_{\phi_c} \cap [E]_{\phi_a} \cap [E]_{\phi_t}$ is the set of events in the log having that execution context.

3.2.2 Quality Measures for Execution Contexts

Given an event log, any categorization rules — as long as they fulfill the requirement (Definition 3.3) — can be proposed for defining types (recall the two alternatives of grouping customer types to define case types). This means that many candidate sets of execution contexts may exist for the same input event log. In this section, we discuss how to measure the quality of execution contexts learned from event logs.

Execution contexts can be applied to characterize resource behavior that concerns certain process execution features determined by the specialization of work, a.k.a. division of labor [29]. On the one hand, when specialization is low in a process, resources tend to be interchangeable when performing in process execution, and events that they originated tend to be similar. On the other hand, when specialization is high, resources are limited to undertaking specific kinds of tasks, as exhibited by the differences among their originated events. This idea motivates us to consider the

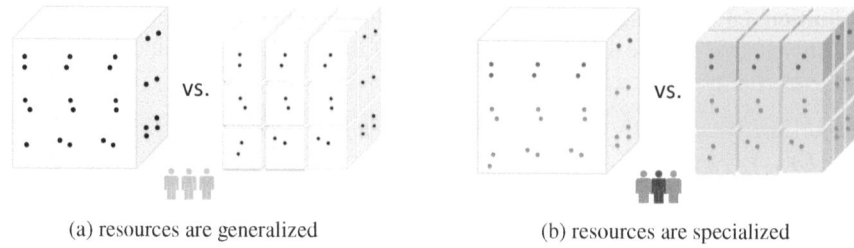

(a) resources are generalized   (b) resources are specialized

Fig. 3.1: High-quality execution contexts should reflect the specialization of resources, so it is desirable to use a small number of dedicated execution contexts (cells) to characterize resource behavior recorded in events.

following criteria for a set of *high quality* execution contexts: (i) events in the same execution contexts should be originated by few resources, and (ii) events originated by the same resource should be partitioned into few execution contexts.

Figure 3.1 illustrates the idea. When resources are considered generalized due to low specialization of work, a small number of execution contexts should be sufficient, therefore the left execution contexts in Figure 3.1a are better. When resources are highly specialized, it is desirable to have dedicated execution contexts for each of them to capture their specific characteristics, therefore the right execution contexts in Figure 3.1b are better. Following this idea, we define two quality measures, namely *impurity* and *dispersal*.

Impurity measures the extent to which the same execution context contains events originated by different resources (Definition 3.5). High-quality execution contexts have low impurity, i.e., an execution context can specifically characterize the behavior of a limited number of resources. This is built upon the existing measure of entropy in data mining [42].

Definition 3.5 (Impurity) Let $EL = (E, Att, \pi)$ be an event log and CO a set of execution contexts,

$$Imp(EL, CO) = \frac{1}{\sum_{r \in \mathcal{R}} p_r \log_2 p_r} \sum_{co \in CO} \left(\frac{|[E]_\mathcal{R} \cap [E]_{co}|}{|[E]_\mathcal{R}|} \times \sum_{r \in \mathcal{R}} p_{r,co} \log_2 p_{r,co} \right)$$

is the impurity of CO with regard to EL, where

$$p_r = \frac{|[E]_r|}{|[E]_\mathcal{R}|}, \quad p_{r,co} = \frac{|[E]_r \cap [E]_{co}|}{|[E]_\mathcal{R} \cap [E]_{co}|}$$

are the relative frequency of events originated by a resource r in terms of the entire log and an execution context co, respectively.

Impurity yields a value in $[0, 1]$. If there is only one execution context for all events in a log, then $Imp(EL, CO) = 1$.

3.2 Problem Modeling

Dispersal measures the extent to which events originated by the same resource disperse across different execution contexts (Definition 3.6), and yields a value in [0, 1]. High-quality execution contexts have low dispersal, i.e., the behavior of an individual resource can be characterized by a limited number of execution contexts.

Definition 3.6 (Dispersal) Given an event log EL and a set of execution contexts CO,

$$Dis(EL, CO) = \sum_{r \in \mathcal{R}} \left(\frac{|[E]_r|}{|[E]_\mathcal{R}|} \times \frac{\sum_{e_1,e_2 \in [E]_r} d_{CO}(e_1, e_2)}{\binom{|[E]_r|}{2}} \right)$$

is the dispersal of CO with regard to EL, where $d_{CO}(e_1, e_2)$ denotes the distance between any two events e_1, e_2.

Let us define the distance between events considering the given set of execution contexts CO. Any event $e \in E$ corresponds to a unique execution context $co^e = (ct^e, at^e, tt^e) \in CO$, for which $e \in [E]_{co^e}$. Then, for any two events $e_1, e_2 \in E$, we define the distance between them using their corresponding execution contexts $co^{e_1} = (ct^{e_1}, at^{e_1}, tt^{e_1})$ and $co^{e_2} = (ct^{e_2}, at^{e_2}, tt^{e_2})$, that is,

$$d_{CO}(e_1, e_2) = \frac{[ct^{e_1} \not\equiv ct^{e_2}] + [at^{e_1} \not\equiv at^{e_2}] + [tt^{e_1} \not\equiv tt^{e_2}]}{ndim}, \quad (3.1)$$

where $[\varphi]$ is the Iverson bracket that returns 1 if a boolean formula φ holds and 0 otherwise, and $ndim \in \{1, 2, 3\}$ is the number of process dimensions considered in a set of execution contexts. By default, we let $ndim = 3$. However, it is possible that there are not any types defined on a dimension. For example, the case dimension can be omitted if, for any $(ct, at, tt) \in CO$, $ct = \phi_{true}$, and thus we have $ndim = 2$.

Specifically, if there is only one execution context for all events in a log, then $Dis(EL, CO) = 0$.

We can combine impurity and dispersal into a single score measuring the overall quality of execution contexts. In this research, we use the harmonic mean as follows.

$$score(EL, CO) = \frac{2}{(1 - Imp(EL, CO))^{-1} + (1 - Dis(EL, CO))^{-1}}. \quad (3.2)$$

Note that we subtract impurity and dispersal from 1 so that a higher score indicates better execution context quality.

3.2.3 Problem Statement

With a viable representation of execution contexts and measures to assess their quality, we now formalize the problem of learning execution contexts: Given an event log, derive three sets of categorization rules that define case types, activity types, and time types, respectively, such that the resulting execution contexts have high quality with respect to the input log.

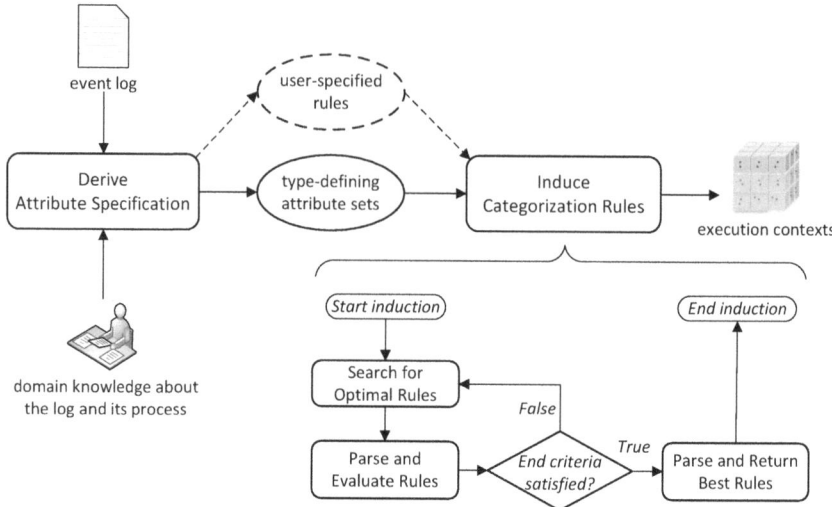

Fig. 3.2: An illustration of the proposed approach to learning execution contexts from an event log.

3.3 Problem Solution

To address the stated problem, we propose an approach that aims at iteratively searching for the optimal categorization rules. Figure 3.2 illustrates the approach. Below, we elaborate on each of the steps involved.

3.3.1 Deriving Attribute Specification

Input to the approach consists of an event log and domain knowledge from the user. First, an *attribute specification* (Definition 3.7) is derived to capture user domain knowledge about the event attributes in the log. An attribute specification comprises (i) X_{case}^{EL}, X_{act}^{EL}, X_{time}^{EL}, which are three sets of type-defining attributes (see Definition 3.1) regarding case types, activity types, and time types; and *optionally* (ii) Λ, which is a function that maps type-defining attributes onto categorization rules supplied by users to capture any existing categorization of attribute values. If no rules are supplied by users for any event attribute x, then $\Lambda(x) = \emptyset$.

All type-defining attributes in an attribute specification are expected to be discrete attributes. For continuous attributes stored in the input event log, their numeric values are expected to have been discretized, i.e., replaced by categorical or interval labels. For example, attribute *cost* records many positive real numbers that can be discretized by intervals like [0, 10000) and [10000, ∞). Data discretization is common in data preprocessing and can be approached in many ways such as histogram analysis [42].

3.3 Problem Solution

Users can specify whether supplied categorization rules are *normative* or *informative*. If a rule is normative, it indicates that certain values of an event attribute have to be categorized together for any rules involving this attribute. Otherwise, it is informative and can be used in the subsequent search as heuristics, but is not enforced.

Definition 3.7 (Attribute Specification) Let $EL = (E, Att, \pi)$ be an event log and let Φ be the set of all possible categorization rules defined on Att. $S = (X_{case}^{EL}, X_{act}^{EL}, X_{time}^{EL}, \Lambda)$ is an attribute specification on EL for learning categorization rules. $X_{case}^{EL} \subseteq Att$, $X_{act}^{EL} \subseteq Att$, and $X_{time}^{EL} \subseteq Att$ are three disjoint, non-empty sets of type-defining attributes. $\Lambda : Att \nrightarrow \mathcal{P}(\Phi)$ defines a set of categorization rules for the event attributes in $X_{case}^{EL} \cup X_{act}^{EL} \cup X_{time}^{EL}$.

An attribute specification informs how attribute values should be handled in the following step of categorization rules induction. Below, we introduce how that can be approached by following two alternative schemes, *decision tree learning* and *simulated annealing*.

Inducing Rules via Decision Tree Learning

In decision tree learning, a dataset of multivariate data tuples is iteratively partitioned into smaller subsets by deriving splitting rules regarding data attributes. The output can be represented in a tree structure, where tree nodes hold the subsets of the input data and branches record the disjunctive splitting rules used to obtain the subsets. Decision tree learning is a common solution for classification tasks, where splitting rules are often derived following a greedy heuristic that minimizes the information needed to classify data tuples.

Searching for Optimal Rules

Decision tree learning provides an intuitive means for solving the problem of learning execution contexts — to derive categorization rules that result in the partitioning of events by event attributes. Note that the learning of execution contexts imposes unique challenges compared to conventional decision tree learning for predictive tasks like classification and regression: (i) we require splitting rules extracted from a decision tree to be categorization rules which can be used for defining types; and (ii) the goal of learning is to derive high-quality execution contexts instead of training a predictive model that aims at high accuracy. To address these issues, we need to customize the generic decision tree learning method.

For the first issue regarding categorization rules, we choose to construct *Oblivious Decision Trees* (ODT) [54]. An ODT is different from a conventional decision tree in such a way that an ODT's nodes at the same level are constructed by splitting rules based on the *same* data attribute. For any two leaf nodes on an ODT, if we project their data subsets onto a split attribute, then the two projected sets are either disjoint or identical. This feature ensures that a learned ODT can be used to produce categorization rules for defining types.

For the second issue regarding the learning goal, we use the harmonic-mean-based quality score (Equation 3.2). Therefore, whenever there exist several sets of categorization rules as candidates, we choose the one that grows the current tree with the highest quality.

Algorithm 1 describes the customized decision tree learning method. It begins with an empty root node that holds all events in a given log (Line 2). At each iteration, it attempts to find the best split, i.e., the best type-defining attribute and the corresponding categorization rules to be applied (Line 4). The selection is done through a sub-procedure FindBestSplit (Lines 15–26). If the result categorization rules are equivalent to the user-supplied rules for that selected attribute, then the user-supplied rules are discarded from future iterations (Line 6). Specifically, if those rules are normative, then the selected attribute is discarded (Lines 7–8) to ensure that user-supplied normative rules are enforced. If the best split can be found, the tree is updated (UpdateTree), i.e., applying the categorization rules to every leaf node in the attempt to grow a subtree therein (Lines 10–11). This ensures that the tree is iteratively constructed as an ODT. The decision tree keeps growing until one of the end criteria is met, i.e., the next best split cannot be found (Lines 12–13) or the height of the decision tree exceeds a preset maximum value (Line 3). After the tree induction halts, events held by the leaf nodes of the tree correspond to the induced categorization rules. We then use the leaf nodes and the split rules derived during tree induction to parse for execution contexts (Line 11). We elaborate on the parsing step in the next section.

FindBestSplit (Lines 15–26) is a sub-procedure that cherry-picks a type-defining attribute and the corresponding categorization rules, i.e., the best split. First, for each type-defining attribute, the user-supplied categorization rules are retrieved, if they exist (Lines 18–19). Otherwise, consider this attribute a generic data attribute and generate sets of candidate rules based on the partition of attribute values (Lines 20–22), for example, randomly choose one from the existing parts and randomly split it into two smaller parts. Then, among all generated rules, select the set that would lead to an expanded tree with the highest quality. In either case, a set of best rules for each type-defining attribute is determined. Finally, return the attribute and its corresponding rules that would lead to the highest quality (Lines 24–25). Note that type-defining attributes are used with replacement when constructing a tree unless there are user-supplied normative rules defined. Therefore, values of a type-defining attribute may be split more than once in tree induction.

EvaluateRules (Lines 27–29) tests a set of rules by applying them to the current tree (i.e., *tree* given as an input) and creating a "test tree" (i.e., *tree'*), which can then be parsed and evaluated. The quality of the execution contexts corresponding to this "test tree" represents the quality of the input set of rules. We elaborate on this in the following section.

The values of dispersal and impurity are expected to show opposite trends as a decision tree grows. Initially, all events are placed together. Hence, dispersal is 0

3.3 Problem Solution

Algorithm 1: Applying decision tree learning to induce categorization rules

input : $EL = (E, Att, \pi)$, an event log;
$S = (X_{case}^{EL}, X_{act}^{EL}, X_{time}^{EL}, \Lambda)$, an attribute specification;
H, the maximum height of the tree to be induced
output a decision tree encoding the induced categorization rules

1 $X \leftarrow X_{case}^{EL} \cup X_{act}^{EL} \cup X_{time}^{EL}$
2 Initialize *tree* as a single root node holding all events E
3 **for** $h \leftarrow 1$ **to** H **do**
4 $\quad$ /* find an attribute and corresponding rules */
$\quad$ $x, \Phi \leftarrow$ FindBestSplit(*tree*, X, Λ)
5 $\quad$ **if** $\Phi = \Lambda(x)$ **then**
6 $\quad\quad$ $\Lambda(x) \leftarrow \varnothing$
7 $\quad\quad$ **if** Φ *is normative* **then**
8 $\quad\quad\quad$ $X \leftarrow X \setminus \{x\}$
9 $\quad$ **if** $\Phi \neq \varnothing$ **then**
10 $\quad\quad$ **for** *leaf* $\in$ *tree* **do**
11 $\quad\quad\quad$ *tree* $\leftarrow$ UpdateTree(*tree*, Φ)
12 $\quad$ **else**
13 $\quad\quad$ **break**
14 **return** *tree*
15 **Function** FindBestSplit(*tree*, X, Λ):
16 $\quad$ $x^* \leftarrow \varnothing$; $\Phi^* \leftarrow \varnothing$; $Q^* \leftarrow$ ParseEvaluate(*tree*)
$\quad$ /* test on all attributes */
17 $\quad$ **for** $x \in X$ **do**
18 $\quad\quad$ **if** $\Lambda(x) \neq \varnothing$ **then**
$\quad\quad\quad$ /* use user-supplied rules, if exist */
19 $\quad\quad\quad$ $\Phi' \leftarrow \Lambda(x)$
20 $\quad\quad$ **else**
$\quad\quad\quad$ /* otherwise, generate rules randomly */
21 $\quad\quad\quad$ Generate C, a set of rules that partition the values of x
$\quad\quad\quad$ /* select candidate rules that lead to the highest quality */
22 $\quad\quad\quad$ $\Phi' \leftarrow \max_{\Phi \in C}$ (EvaluateRules(*tree*, Φ))
23 $\quad\quad$ $Q' \leftarrow$ EvaluateRules(*tree*, Φ')
$\quad\quad$ /* select attribute that leads to the highest quality */
24 $\quad\quad$ **if** $Q' > Q^*$ **then**
25 $\quad\quad\quad$ $Q^* \leftarrow Q'$; $x^* \leftarrow x$; $\Phi^* \leftarrow \Phi'$
26 $\quad$ **return** x^*, Φ^*
27 **Function** EvaluateRules(*tree*, Φ):
$\quad$ /* test a set of candidate rules */
28 $\quad$ *tree'* $\leftarrow$ UpdateTree(*tree*, Φ')
29 $\quad$ **return** ParseEvaluate(*tree'*)

while impurity is 1. As the decision tree grows, the number of leaf nodes increases (so is the number of their corresponding execution contexts), which leads to an increase in dispersal and a decrease in impurity.

Parsing and Evaluating Rules

As mentioned, the parsing and evaluation of categorization rules happen both when we need to evaluate intermediate solutions and when we need to obtain the optimal final result after the search terminates. In the context of decision tree learning, we first follow the conventional way of rule extraction from a decision tree. That is, for each path from the root to a leaf node, a decision rule is formed as a logical conjunction of all the rules recorded along the path. Then, for every decision rule obtained, we use the attribute specification as a reference to determine which part of the decision rule is related to case types, activity types, or time types, respectively. Formally, every such decision rule ϕ can be written as a conjunction ($\phi_c \wedge \phi_a \wedge \phi_t$), where any of ϕ_c, ϕ_a, ϕ_t can be a default rule (ϕ_{true}) if no type-defining attributes are included for any of the core event attributes.

As such, we will be able to transform a decision rule related to a leaf node of a decision tree into an execution context $co = (\phi_c, \phi_a, \phi_t)$. A set of execution contexts CO is obtained by parsing the categorization rules for all leaf nodes. Then, we can calculate the impurity (Definition 3.5) and dispersal (Definition 3.6) of CO to evaluate its quality.

Let us analyze the time complexity of applying decision tree learning to induce categorization rules. We refer to Algorithm 1 and adopt the worst-case estimate, assuming that (i) at each iteration, any split results in all leaf nodes being split into two; (ii) the tree induction continues until the tree grows to the maximum height allowed, or all possible splits are exhausted; and (iii) no user-supplied rule is available for any type-defining attribute. Note that the last condition means that for any attribute, rules have to be randomly generated (Lines 20-22), leading to a set of candidates $|C|$ to be parsed an evaluated. The number of candidates is at most $2^{n-1} - 1$, i.e., no split has been applied to an attribute with n distinct values. Clearly, in practice, it is infeasible to enumerate all possible candidates. Hence, we assume $|C|$ is bounded by some given neighborhood size N, i.e., $|C| = \min(N, 2^{n-1} - 1)$.

Denote the number of events in the input log by $|E|$ and the number of distinct resources by $R = |rng(\pi_{res})|$. Let M be the number of loops for tree induction (the loop in Lines 3–13). By the worst-case assumption, M is bounded by H and the number of loops required to exhaust all possible splits. The latter can be determined based on the number of distinct values of type-defining attributes. Therefore, $M = \min(H, \sum_{x \in X} |rng(\pi_x)| - 1)$. At each iteration $h \in [1, M]$, we denote l the number of leaf nodes — in the worst case, we have $l = 2^h$. FindBestSplit requires looping over $|X|$ attributes. UpdateTree applies a candidate set of rules to create a "test tree" for evaluation. This takes $O(l)$, i.e., every existing node is split into two. Then, the evaluation calculates impurity and dispersal. The calculation of impurity takes $O(lR)$ by Definition 3.5. The calculation of dispersal would take $O\left(|E|^2\right)$ by Definition 3.6 — however this can be alternatively done in $O(l^2R)$ since the pairwise event distance is defined based on the pairwise execution context distance. Therefore, the parsing and evaluation for each candidate take $O(lR + l^2R)$. And the complexity at iteration h is thus $O\left(2^h RN |X| + 4^h RN\right)$.

3.3 Problem Solution

Summing up, the overall time complexity of applying decision tree learning in the worst case is $O\left(2^M RN |X| + 4^M RN\right)$. In other words, the time complexity is linear to the number of resources in the log (R), the neighborhood size (N), and the number of type-defining attributes ($|X|$); and it is exponential to the given maximum tree height and the total number of distinct values of type-defining attributes (M).

The customized decision-tree-based method provides an intuitive solution to inducing categorization rules. This method constructs a set of execution contexts by gradually splitting an event log toward lowering impurity and dispersal. The tree representation may be utilized to understand how execution contexts are derived incrementally. But it has certain limitations. The initial state is always a root node holding all events, and the partition of events will only be modified in a sequential forward manner. Also, the tree induction procedure follows a greedy heuristic, as splits happen only if the expanded tree in the next iteration has better quality compared to the current iteration. Consequently, for some given input, the decision-tree-based method is likely to produce outputs close to the same locally optimal results. To overcome these limitations, we introduce the second solution that applies the simulated annealing algorithm to search for near-global-optimal categorization rules.

3.3.2 Inducing Rules via Simulated Annealing

Simulated Annealing (SA) is an established technique for solving many generic combinatorial optimization problems [79]. SA enables searching for near-global-optimal solutions in a solution space (i.e., the universe of all solutions to the problem at hand) having many poor local optima. The core idea is based on the analogy of metal cooling in thermodynamics [53]. The search strives to find better solutions iteratively but allows moving to worse solutions by some probability, which is initially high and decreases gradually — as the "system temperature" cools down. This way, SA may explore a wider solution space and avoid being trapped in a local optimum. In the meantime, it has fewer parameters to configure, compared to other heuristic techniques such as the Genetic Algorithm. SA has been shown to be a robust and relatively efficient algorithm for solving single-objective as well as multi-objective optimization problems [76].

Search for Optimal Rules

To apply SA, we first need to encode the solutions to the problem, define the objective function and constraints, and configure the search parameters.

Here, a solution to the problem is a set of categorization rules that define execution contexts. Following the definition of categorization rules (see Definition 3.2), we encode solutions as partitions on the values of every type-defining attribute given in an attribute specification. As mentioned, note that all input attributes are expected to be discrete, hence the partitions are finite sets. Formally, such a solution in the form of partitions is a mapping,

$$P: \mathcal{U}_{Att} \nrightarrow \mathcal{P}(\mathcal{P}(\mathcal{U}_{Val})).$$

For any type-defining attribute $x \in X = X^{EL}_{case} \cup X^{EL}_{act} \cup X^{EL}_{time}$, $P(x)$ is a partition of its attribute values.

A mapping P specifies a set of categorization rules that define execution contexts. For $x \in X$, the set of categorization rules specified by P is

$$\Phi_x = \{ \phi(x, p) \mid p \in P(x) \}, \text{ with } \phi(x, p) = \bar{x} \triangleleft \bar{p}.$$

Note that $\phi(x, p)$ is a categorization rule by Definition 3.2. For any core event attribute $d \in D$, a set of types can be defined as the conjunction of the rules for all its related type-defining attributes. Take the activity dimension as an example (that is, $d = act$). Activity types can be defined by the conjunction of rules for attributes in X^{EL}_{act}. Formally, let I be some indexing set for X^{EL}_{act}, then an attribute in X^{EL}_{act} can be indexed by a_i. The set of categorization rules that define activity types is

$$\Phi_{act} = \left\{ \bigwedge_{i \in I} \sigma(i) \,\middle|\, \sigma \in \bigtimes_{i \in I} \Phi_{a_i} \right\},$$

i.e., for each combination of the individual rules ($\sigma(i)$ in the sequence σ) for an attribute, their logical conjunction denotes a rule that defines an activity type. Rules for case types (Φ_{case}) and time types (Φ_{time}) can be determined in the same way. As a result, an encoded solution is translated into three sets of categorization rules that define a set of execution contexts (Definition 3.4).

If normative rules are supplied by a user in the attribute specification, these rules cannot be violated in a solution. Formally, given an attribute specification S, where X is the union of type-defining attributes, and Λ, a function capturing user-supplied categorization rules related to those attributes, if P is a feasible solution, then

$$\nexists x \in X \left[\Lambda(x) \text{ is normative} \land \exists_{\tilde{\phi} \in \Lambda(x)} \left(\forall_{\phi \in \{ \phi(x,p) \mid p \in P(x) \}} \phi \not\equiv \tilde{\phi} \right) \right].$$

In other words, a feasible solution should specify a set of categorization rules that subsume existing user-supplied normative rules for any attribute.

The objective function is defined by the quality score (Equation 3.2) that combines impurity and dispersal. This is consistent with the learning goal setting in the decision-tree-based method.

Algorithm 2 describes the main procedure of applying SA to induce categorization rules. It starts with initializing P, which encodes a partition over the values of every type-defining attribute (Lines 3–9). For any attribute $x \in X$, if there exist user-supplied rules $\Lambda(x)$, the partition is defined exactly as the sets of attribute values expressed in the rules. In addition, if those rules are normative, then this attribute is discarded from the future modification of P. Otherwise, we can initialize a partition ϱ using (i) empty initialization: $\varrho \leftarrow \{rng(\pi_x)\}$, i.e., the trivial partition;

3.3 Problem Solution

or (ii) random initialization: ϱ is a partition randomly sampled from $\mathcal{P}(rng(\pi_a))$. Then, the initial solution is parsed and evaluated (ParseEvaluate, Line 11), which is done by translating it into a set of execution contexts and calculating impurity and dispersal.

Then, an optimal solution can be derived via searching (Lines 12–22). The search starts by taking the initial solution as the current best and setting the initial temperature (Lines 12–13). The following steps then iterate: generating a neighboring solution (Line 16), deciding probabilistically whether to accept the neighboring solution by comparing it to the current one (Lines 17–19), and updating the tracked best solution (Lines 20–21). The temperature gradually decreases during the iteration following the selected cooling schedule (Line 22), until it drops below the minimum temperature allowed, i.e., the end criterion is satisfied. The search halts and returns the best solution found so far as the final solution.

Algorithm 3 describes how a neighboring solution is generated based on the current one. To begin with, an attribute a is randomly selected, and the partition of its values $P(a)$ will be randomly modified. We consider two operators, *split* and *merge*, which have equal probabilities of being applied to $P(a)$ (Line 4).

- A split can be applied to a partition unless all parts are singletons (Lines 10–11). A split is done by randomly selecting a non-singleton part (Line 13) and then choosing two non-empty, proper subsets to substitute it in the partition (Lines 14–17).
- A merge can be applied to a partition unless it is a trivial partition (Lines 20–21). A merge is done by randomly selecting two parts without replacement (Line 23) and then using their union to substitute them in the partition (Lines 23–24).

We discuss how to configure the parameters. Neighborhood size (N) controls the number of neighboring solutions to generate and test at each iteration (Line 15). With a larger N, SA tests more solutions and vice versa. This resembles the set of candidate rules in the decision-tree-based method (cf. Line 21 in Algorithm 1, the set C). The initial system temperature (T_0), the minimum temperature allowed (T_m), and the cooling strategy are specific to SA. Together, these determine the search behavior: when the system temperature is high ($T \rightarrow T_0$), the search tends to explore a wide range of the solution space by accepting worse solutions with high probability; when the temperature is low ($T \rightarrow T_m$), the search tends to move greedily by accepting only better solutions; and the cooling schedule decides how T decreases from T_0 to T_m. Configuring the temperatures and the cooling schedule allows the proposed SA-based method to produce near-global-optimal execution contexts. However, to give a theoretical estimate of the "optimal" parameter values is nontrivial and is nonunique to the problem of learning execution contexts. In applying SA, it is common to empirically determine the temperature parameters and the cooling schedule through experiments on the same dataset (in this case, an event log and an attribute specification). For more details, refer to the dedicated literature [87, 11, 46].

Let us analyze the time complexity of the SA-based method. We focus on the iterative part (Lines 14–22 in Algorithm 2). Denote M as the number of the outer

Algorithm 2: Applying simulated annealing (SA) to induce categorization rules

input : $EL = (E, Att, \pi)$, an event log;
$S = (X_{case}^{EL}, X_{act}^{EL}, X_{time}^{EL}, \Lambda)$, an attribute specification;
N, neighborhood size;
T_0, the initial system temperature;
T_m, the minimum temperature allowed;
A cooling schedule

output P^*, a set of partitions to be parsed for the optimal categorization rules

1 $X \leftarrow X_{case}^{EL} \cup X_{act}^{EL} \cup X_{time}^{EL}$
   // initialize solution and evaluate
2 $P \leftarrow \emptyset$
3 **for** $x \in X$ **do**
4   **if** $\Lambda(x) \neq \emptyset$ **then**
5   Initialize a partition ϱ according to $\Lambda(x)$
6   **if** $\Lambda(x)$ *is normative* **then**
7            $X \leftarrow X \setminus \{x\}$
8   **else**
9   Initialize a partition ϱ on the attribute values $\{ \pi_x(e) \mid e \in E \}$
10     $P \leftarrow P \cup \{(x, \varrho)\}$
11 $Q \leftarrow \mathsf{ParseEvaluate}(P)$
   // search iteratively
12 $P^* \leftarrow P \,;\, Q^* \leftarrow Q$
13 $T \leftarrow T_0$
14 **while** $T \geq T_m$ **do**
15   **for** $i \leftarrow 1$ **to** N **do**
16       $P' \leftarrow \mathsf{GenerateNeighbor}(P, X)$
17       $Q' \leftarrow \mathsf{ParseEvaluate}(P')$
    // determine acceptance of the neighbor solution
18   **if** $Q' > Q \lor \mathsf{Random}(0, 1) < e^{\frac{Q'-Q}{T}}$ **then**
19          $P \leftarrow P', Q \leftarrow Q'$
    // update the best solution if current is better
20   **if** $Q > Q^*$ **then**
21          $P^* \leftarrow P, Q^* \leftarrow Q$
22   Decrease T according to the cooling schedule
23 **return** P^*

while-loops. If the selected cooling schedule is a static schedule [11], then M can be determined by T_0 and T_m. For example, with the widely-used exponential schedule [53], $T_k = \alpha^k T_0$, we have $M = \lceil \log_\alpha (T_m/T_0) \rceil$. To derive M in the cases of dynamic schedules is complex and should be done by consulting the literature on analyzing the complexity of SA.

The inner for-loop (Lines 15–21) is specific to the problem of learning execution contexts. At some temperature T, N neighbors are generated and tested. For each neighbor, ParseEvaluate takes $O\left(lR + l^2 R\right)$, where l is the number of execution contexts corresponding to the solution at the current step and R is the number of distinct resources in the input log. Note that this is the same as the parsing and

3.3 Problem Solution

Algorithm 3: Generating a neighbor solution via splitting or merging an existing partition on the values of an event attribute

input : P, a set of partitions encoding a solution;
X, a set of event attributes
output P', a set of partitions encoding a neighbor solution

1 **Function** GenerateNeighbor(P, X):
2   Select an attribute $x \leftarrow$ Sample(X)
        // modify the partition of the values of x
3      $P' \leftarrow P$
4   **if** Sample($[0, 1)$) < 0.5 **then**
5          $P' \leftarrow P' \oplus \{(x, \text{Split}(P(x)))\}$
6   **else**
7          $P' \leftarrow P' \oplus \{(x, \text{Merge}(P(x)))\}$
8   **return** P'

9 **Function** Split(ϱ):
        // modify by randomly splitting one part into two
10   **if** $\{s \in \varrho \mid |s| > 1\} = \emptyset$ **then**
           // all parts are singletons
11          $\varrho' \leftarrow \varrho$
12   **else**
13   Select a part $p \leftarrow$ Sample($\{s \in \varrho \mid |s| > 1\}$)
14          $\varrho' \leftarrow \varrho \setminus \{p\}$
           // split the part into two
15   Select a subset $q \leftarrow$ Sample($\mathcal{P}(p) \setminus \{\emptyset, p\}$)
16          $p \leftarrow p \setminus q$
17          $\varrho' \leftarrow \varrho' \cup \{p, q\}$
18   **return** ϱ'

19 **Function** Merge(ϱ):
        // modify by randomly merging two parts
20   **if** $|\varrho| = 1$ **then**
           // trivial partition
21          $\varrho' \leftarrow \varrho$
22   **else**
           // select two parts without replacement
23          $p \leftarrow$ Sample(ϱ) ; $q \leftarrow$ Sample($\varrho \setminus \{p\}$)
           // merge two parts
24          $\varrho' \leftarrow \varrho \setminus \{p, q\} \cup \{p \cup q\}$
25   **return** ϱ'

evaluation in the decision-tree-based method. Due to the probabilistic nature of SA in accepting solutions, it is challenging to determine l. Here, we consider an upper bound for the worst-case estimate, $l \leq |\{\pi(e)\!\restriction_X \mid e \in E\}|$. This upper bound states that, for any solution, it encodes a set of partitions that are not finer than the partitions constructed from enumerating every *observed* combination of distinct type-defining attribute values in the log. Determining the acceptance of neighbor solutions takes $O(1)$.

The overall complexity is therefore $O\left(MRNl^2\right)$, which is linear to the number of outer loops (M, determined by the configuration of temperatures and cooling schedule), the number of resources (R), and the set neighborhood size (N); and it is quadratic to the number of unique combinations of distinct type-defining attribute values observed in the log (i.e., the upper bound of l).

Parsing and Evaluating Rules

This step was discussed in the previous introduction to solution encoding — a set of partitions corresponds to a set of categorization rules, which can then be evaluated.

The SA-based method is an improved solution to inducing categorization rules. It has several advantages over the foregoing customized decision-tree-based method: (i) allowing random initialization, (ii) using both the split and merge operators, and (iii) adopting an effective heuristic to avoid local optima. As such, applying the SA-based method can lead to finding a set of execution contexts with better quality compared to those produced by the application of the decision-tree-based method. The SA-based method has its limitations. Its application requires adjusting the temperature parameters and choosing a cooling schedule — as mentioned, in practice, these are often achieved through empirical tests and hence may cost additional time and effort to obtain high-quality results.

3.4 Evaluation

We implemented the proposed approach and evaluated it through experiments. In this section, we first report on the experiment datasets and explain the experiment setup. We then present the experiment results and findings.

3.4.1 Event Log Datasets

In total, five datasets were used for evaluation [33, 34, 35, 63, 25]. They contain event logs recording business processes in real-world organizations from three industry sectors. All datasets are deposited in an online repository maintained by 4TU.ResearchData[10] and are made publicly available for use by academic research. Three datasets [33, 34, 35] were originally released for the Business Process Intelligence Challenge (BPIC)[11], where real-world organizations share their process execution data and propose business questions to be addressed through the application of process mining and other data analytics approaches. Two other sets [63, 25] were released as case study data in published research [64, 24]. We selected these datasets based on the following criteria:

---

[10] 4TU.ResearchData: https://data.4tu.nl/

[11] Business Process Intelligence Challenge: https://www.tf-pm.org/competitions-awards/bpi-challenge

3.4 Evaluation

- Data should record a minimum number of human resources, so it is possible to perform analyses regarding their organizational groupings. In our evaluation, we use 10 as the minimum number of resources.
- Data should record at least one event attribute in addition to the core ones, i.e., case identifier, activity name, timestamps, and resource identifier, that can be qualified as a type-defining attribute for learning execution contexts from event logs;
- Data should be enclosed with metadata of the recorded attributes so that any mining and analysis results can be interpreted in a meaningful way.

Table 3.1 summarizes the basic characteristics of the selected datasets. Names of the datasets are shortened for conciseness. Below, we introduce each event log dataset, covering the process, the organization, and the recorded event logs. Some of these datasets are also used in later chapters of this book.

Table 3.1: A summary of the characteristics of the selected event log datasets.

Log	Industry	Timespan (months)	#cases	#events	#activities	#resources
bpic15	Government administration	57.1	5649	262628	496	72
bpic17	Banking and finance	13.3	31509	1202267	26	149
bpic18	Government administration	45.2	43809	2514266	41	165
sepsis	Health services	19.2	1050	15214	16	25
wabo	Government administration	16.0	1434	8577	27	48

BPIC'15 Log [33]

Log bpic15 originates from the five event logs recording a building permit application process in five Dutch municipalities from 2009 to 2015. The data was released for BPIC 2015, with a set of business questions that aimed at comparing the differences between municipalities in terms of the workforce and their performance.

The process can be considered as mostly identical across the municipalities [33]. Hence, we generated a single log by merging the five event logs and preserving unique cases and municipality identities. Note that the process contains many activities (496), but they can be grouped into subprocesses and further into different phases. Each resource recorded in the log refers to one of the 72 employees in the municipalities. Most of them worked for a single municipality during the period of data recording, while some performed tasks for different municipalities. Each case in the log is a building permit application, for which we can determine its permit type (e.g., construction or destruction) and its responsible resource based on the case attributes.

We derived three additional attributes based on the original data and its description. Case attribute "case:parts Bouw" is a Boolean attribute indicating whether an application is related to a construction permit. Event attributes "subprocess" and "phase" were derived based on the grouping of process activities, which is indicated by the values of an event attribute "action code" [33].

BPIC'17 Log [34]

Log bpic17 records a loan application process in a financial institute. The data was collected from the organization's workflow system and contains all applications filed in 2016 and their handling up to February 2017.

For a loan application, case attributes in the data record information such as the loan goal and application type. Also, there may be multiple offers granted for a single application. Hence, cases contain three types of events, i.e., application state changes, offer state changes, and workflow events. Note that in bpic17, information about the lifecycle of activities is recorded through the transaction type attribute [4]. This means, for an activity instance performed in process execution, the log may have recorded more than one event pointing out its start and completion, and intermediate states such as being assigned to a resource. In this research, we consider only the completion of activity instances and take resources who originated the completion events as performers of process activities. In total, there were 149 resources involved in processing 31509 loan applications.

BPIC'18 Log [35]

Log bpic18 contains execution data of a process handling applications for direct payments to German farmers from the European Agricultural Guarantee Fund. The data was extracted from an enterprise system deployed in four local departments, recording the process execution from 2014 to 2018.

The application-handling process can be understood based on different document types [35]. An application was concerned with several documents containing various types of information required to assess the application, e.g., inspection results and annual payments. A document was handled by some staff in a department, following different subprocesses. Activities in these subprocesses represent the states of the documents after being handled by the staff. In the event log, an application corresponds to a case. An event within the case records a resource (staff member) handling some document related to that application in a subprocess. In other words, an activity instance in a case should be identified based on combining the document type, the subprocess, and the state of the document. In total, 165 resources were involved in the process execution, including non-human-resources such as the workflow system distributing the documents ("document processing automaton"). Many case attributes are available in this log, for example, the application type, the type of penalties applied to applications, and the risk assessment results of applications.

Specifically, the data description [35] notes that there were some major changes to the document types used in the process, due to changes in regulations or technical implementation. To keep a consistent view of the activity instances in the log, we will focus on a subset of data where cases started on or after 2017-01-01 — no further change to document types took place after this point. The selected subset contains 14507 cases, which comprises 33% of all cases in the original dataset.

Sepsis Cases Log [63]

Log sepsis records a healthcare process from a regional hospital in the Netherlands. The data was originally recorded by the hospital's ERP system from the year 2013 to 2015 and was collected and anonymized for a case study by Mannhardt and Blinde [64].

The healthcare process represents the pathway of sepsis patients through the hospital from admission to discharge. This process has 16 activities, which can be grouped into six phases, i.e., registration and triaging, admission or transfer, measurement, giving infusions, discharge, and dealing with returning patients. Each case in the event log records a patient's trajectory, and the events contain data related to the activities performed by the clinical groups in the hospital to care for the patient. In total, the log records 25 clinical groups (resources). Additionally, the log also includes 25 case attributes sourced from the triage documents, which contain checklists filled in for patients when they were admitted to the hospital. Several business questions were identified and investigated, regarding (i) the conformance to medical guidelines for the treatment of sepsis, (ii) the analysis of specific patient trajectories, e.g., admission to normal care and admission to intensive care, and (iii) trajectories of patients returning within 28 days. More details can be found in the data description [63] and the article reporting the case study [64].

We derived two additional attributes based on the original data and its description. A case attribute "case:returning" takes Boolean values indicating whether a patient is a returning patient. An event attribute "phase" takes categorical values showing the phase of the process activity recorded by an event.

WABO Receipt Phase Log [25]

Log wabo records the receipt phase of a building permit process performed in a municipality from 2010 to 2012. The data was collected as part of the Configurable Services for Local Governments (CoSeLoG) research project and was reported in the doctoral thesis of Buijs [24].

The process consists of 27 activities, mainly concerned with the municipality handling documents relevant to the receipt of building permits, e.g., creating, checking, and adjusting documents. The event log records 1434 cases of the process that involved a total of 48 individual workers performing the process activities. Several case attributes are recorded. However, only a limited number of them have metadata available. This is likely due to the fact that the original research [24] focused on the control-flow perspective of the process and did not report on the use of those attributes.

3.4.2 Experiment Setup

The purpose of the experiments is two-fold: (i) to test the feasibility of our approach in solving the problem of learning execution contexts, and (ii) to compare the effectiveness and efficiency of the decision-tree-based and the SA-based method.

To this end, we need to preprocess the original event logs before applying the proposed approach. This includes specifying type-defining attributes for the three core process dimensions (case, activity, and time) and applying suitable filters to select relevant data.

Specifying type-defining attributes

We referred to metadata in the dataset descriptions to determine if an attribute is related to any possible definition of types on the core process execution dimensions. When there is not a suitable attribute for a dimension, we applied the following *default* setting: For the activity dimension, we used the activity label. Note that it can be a type-defining attribute by itself as per Definition 3.1. For the time dimension, we derived "month" and "weekday" from the date component of the original timestamps. In the following, we explain the type-defining attribute selection for each dataset.

- bpic15: For case types, we used two attributes. Attribute "case:Responsible actor" is the identity of the responsible resource, and "case:parts Bouw" is a derived attribute indicating whether a case is relevant to construction. For activity types, we used the "subprocess" and "phase" attributes derived.
- bpic17: For case types, we used two attributes — "case:Loan Goal" records the reason used by customers when applying for the loan; "case:Application Type" records the type of application, e.g., if it is applying for a "New credit" or "Limit raise".
- bpic18: For case types, we used 37 attributes. Attribute "case:department" records the department handling the case. There are three case attributes indicating the application type, i.e., whether it is an application for redistributive payment, the small farmer scheme, or the young farmer scheme. Furthermore, there are 30 Boolean attributes indicating the types of penalties. Lastly, there are two case attributes indicating whether the case was selected for inspection and one attribute "case:rejection" indicating whether the case was entirely rejected. For activity types, we used attribute "doctype" (document type) and the attribute indicating the state of a document.
- sepsis: For case types, we used 12 case attributes, among those 11 are related to the type of clinical tests ordered for the patients. The names of these attributes all start with a prefix "Diagnostic", e.g., "DiagnosticBlood" is a selected attribute indicating whether a blood test was ordered. The other one is related to whether a patient is a returning patient (i.e., the derived attribute "case:returning").
- wabo: For case types, we used two attributes: "case:channel" represents the five communication channels used by the customers when applying for permits, and

3.4 Evaluation

"case:department" suggests the expertise demanded to handle the permit (which can be "General", "Expert", or "Customer contact").

Filtering relevant data

With the selected type-defining attributes, we first filter out cases and events that record a null value for any type-defining attribute. Also, we neglect meaningless resource identifiers, such as "?", "test", "n/a". Specifically, as aforementioned, for bpic17 we used only events that record the completion of process activity instances, and for bpic18 we focused on events recorded for the "main" and "application" subprocesses since the start of 2017. As a result, we obtain the preprocessed event logs. Table 3.2 reports their statistics.

Table 3.2: A summary of the selected event log datasets after preprocessing.

Log	#cases	#events	#resources	#type-defining attributes	#distinct type-defining value combinations observed
bpic15	5641	262194	72	6	34969
bpic17	31509	475306	144	5	25904
bpic18	14507	341981	107	41	15221
sepsis	995	13943	25	15	8030
wabo	1434	8570	46	5	1249

Configuring the methods

We discuss the configuration applied to compare the effectiveness of the two proposed methods based on decision tree learning (hereby referred to as tree-based) and SA (hereby referred to as SA-based).

First, we set the SA-based method to use the empty initialization and the widely adopted exponential cooling schedule with a decreasing rate of 0.95. We set the initial and the minimum temperature to 20 and 3×10^{-4}, such that there is a 95% probability of accepting a worst possible neighboring solution at the beginning of the search (i.e., with a quality score difference of 1), and a 5% probability of accepting a neighboring solution of similar quality (i.e., with a score difference smaller than 10^{-3}). We can therefore calculate the number of total iterations as 217 (i.e., $\lceil \log_{0.95}(20/(3 \times 10^{-4})) \rceil$). We set the maximum height parameter to the same number for the tree-based method. Similarly, we set the number of candidate rules of the tree-based method to 1 and the neighborhood size parameter of the SA-based method to the number of type-defining attributes specified for the input log. In summary, the configuration above sets (i) the same initialization, (ii) the same number of total iterations, and (iii) the same neighborhood size, and thus ensures a fair comparison between the two methods.

Both the tree-based and the SA-based method involve random sampling when inducing rules. To avoid arbitrariness in the results, we ran each of the methods 10 times on the same dataset with the same configuration.

In addition, we included a baseline in the comparative evaluation. For an event log, we construct the baseline execution contexts by enumerating the distinct combinations of type-defining attribute values observed in the log. For example, for bpic15, the baseline has 34969 execution contexts with 45 activity types, each corresponds to a unique "phase" in a "subprocess". These baselines represent the results of manually specifying execution contexts without clear prior information or learning from the discriminative information in the logs.

3.4.3 Evaluation against the Baselines

We obtained a total of 100 solutions in the experiments (10 solutions per method per dataset). First, we compared the worst solutions against the baselines to demonstrate the effectiveness of the proposed methods for learning execution contexts. In the analyses below, we consider the use of type-defining attributes, resultant execution context size (number of execution contexts), and quality (impurity, dispersal, and score).

Table 3.3 shows the worst execution contexts learned from the datasets, compared with the baselines. Across all datasets, we can see that the learned execution contexts are smaller in size. Many of those have less than 5% the size of the baselines. These observations, combined with the number of type-defining attributes used to define types, indicate that the proposed methods were able to pick from the given type-defining attributes. Furthermore, the categorization rules learned by the proposed methods were able to group the attribute values. An example is bpic17: the baseline and the SA-based results have the same number of attributes used, yet the latter has a smaller size due to the grouping of attribute values to define rules.

In the meantime, the learned execution contexts — even when they are the worst solutions — still achieved improved quality compared to the baselines. Note that the learned execution contexts are expected to have a higher impurity. This is because the baselines correspond to the most fine-grained set of execution contexts, and hence events in an execution context are less likely to be originated by various resources. On the contrary, the learned execution contexts have much lower dispersal due to the reduced use of type-defining attributes to capture resource specialization. This contributes to the better overall quality of the learned execution contexts, as indicated by the quality score. Specifically, the results from log wabo are less promising, which may be due to resources in the corresponding building permit process being more generalized [88].

3.4 Evaluation

Table 3.3: Comparing the worst solutions produced by the proposed learning execution contexts methods and the baselines.

Log	Method	Use of TD attributes[1]			Size[2]	Quality		
		case	activity	time		impurity	dispersal	score
bpic15	baseline	2	2	2	34969	0.154	0.827	0.287
	tree-based	2	2	1	1255 (-96%)	0.378	0.476	0.569 (+98%)
	SA-based	1	2	1	794 (-98%)	0.421	0.426	0.577 (+101%)
bpic17	baseline	2	1	2	25904	0.508	0.870	0.206
	tree-based	2	1	1	674 (-97%)	0.686	0.600	0.352 (+71%)
	SA-based	2	1	2	6607 (-74%)	0.701	0.669	0.314 (+53%)
bpic18	baseline	37	2	2	15221	0.064	0.689	0.467
	tree-based	4	2	1	228 (-99%)	0.237	0.334	0.712 (+52%)
	SA-based	5	2	2	311 (-98%)	0.275	0.218	0.752 (+61%)
sepsis	baseline	12	1	2	8030	0.033	0.820	0.304
	tree-based	6	1	1	360 (-96%)	0.157	0.552	0.585 (+93%)
	SA-based	1	1	0	15 (-99%)	0.286	0.080	0.804 (+165%)
wabo	baseline	2	1	2	1249	0.490	0.655	0.411
	tree-based	2	1	2	578 (-54%)	0.569	0.549	0.440 (+7%)
	SA-based	1	1	2	648 (-48%)	0.566	0.567	0.433 (+5%)

[1] Number of type-defining attributes used to define execution contexts (per process dimension)
[2] Number of execution contexts

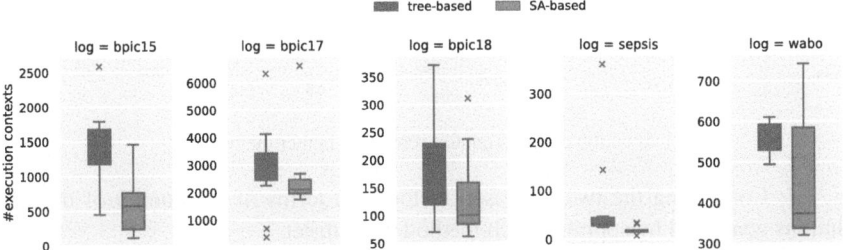

Fig. 3.3: Comparing the two proposed methods in terms of the size (number of execution contexts) of the 10 solutions generated by applying each method per dataset.

3.4.4 Evaluation between tree-based and SA-based

We now proceed to a detailed comparison between tree-based and SA-based. For each dataset, we compare the size (number of execution contexts) and quality of the 10 solutions generated by applying each method.

Figure 3.3 shows the comparison of solution size based on the number of execution contexts. We observe that the SA-based method produced smaller-sized solutions, which means the resultant execution contexts should be simpler.

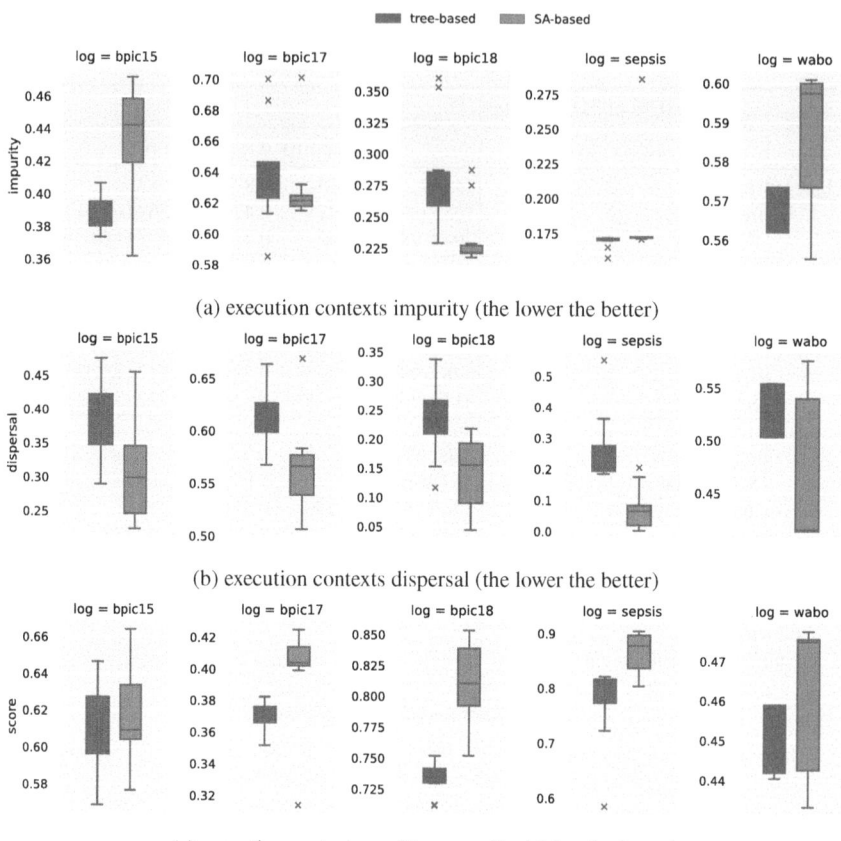

Fig. 3.4: Comparing the two proposed methods in terms of the quality of the 10 solutions generated by applying each method per dataset.

Figure 3.4 illustrates the comparison of solution quality. In terms of impurity (Figure 3.4a), the two methods seem comparable — in bpic15 and wabo, tree-based solutions are better (with lower impurity); in bpic17 and bpic18, SA-based outperformed tree-based; in sepsis, the results are similar. In the meantime, observe that the SA-based solutions usually have lower dispersal (see Figure 3.4b). This, combined with our previous observation about solution size, shows that using SA-based produces execution contexts capable of capturing resource specialization more compactly, compared to tree-based. The reason is that the SA-based method is designed to avoid inducing rules in a sequential forward manner as the tree-based method does. With both the split and merge operators, SA-based can avoid solutions that contain overly fine-grained partitions.

3.4 Evaluation

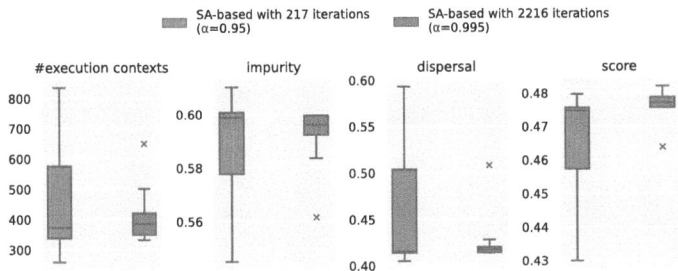

Fig. 3.5: Comparing the solutions obtained by using different numbers of total iterations when applying SA-based on log wabo (additional experiment).

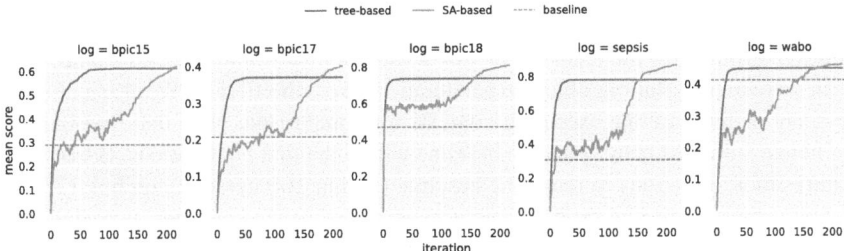

Fig. 3.6: Comparing the two proposed methods in terms of the mean score of solutions obtained per iteration.

In terms of the overall quality score, we can see that the SA-based method can produce solutions with higher (bpic17, bpic18, and sepsis) or at least comparable quality (bpic15 and wabo).

An observation across different datasets and measures is that the SA-based method is less stable than the tree-based method, as implied by the larger interquartile ranges. A possible reason is that the SA-based method explores the solution space more extensively and is therefore more likely to follow various search paths across different runs. Consequently, the final solutions tend to vary, especially when the temperature decreases overly fast and the number of total iterations is not sufficient. To verify this conjecture, we conducted an additional experiment on log wabo, running SA-based 100 times with different temperature decreasing rates (α): 0.95 vs. 0.995. The former was used in the original configuration, resulting in 217 total iterations; the latter is a slower cooling schedule, resulting in 2216 total iterations. Figure 3.5 shows the comparison of the 200 solutions regarding their size and quality. Note that with the slower cooling schedule — and thus more iterations — the results are more stable. They are also generally better in terms of having a smaller size and higher quality.

We also wish to understand how tree-based and SA-based compare with regard to their efficiency in obtaining quality solutions. To this end, we looked at the

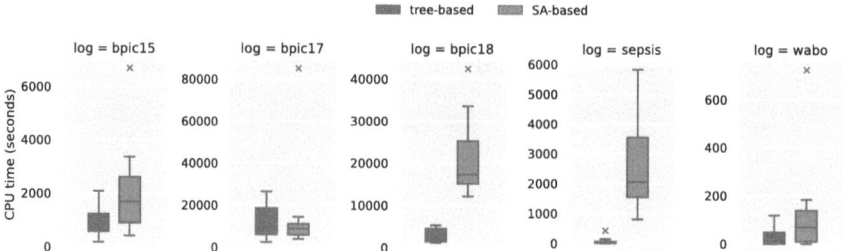

Fig. 3.7: Comparing the two proposed methods in terms of efficiency, measured by CPU time in seconds.

relationship between the score of intermediate solutions and the search iterations. Figure 3.6 illustrates the results. For tree-based and SA-based, we calculated the mean score of the solutions obtained per iteration across the 10 runs. We also included the score of the baseline execution contexts for comparison. Note that tree-based can obtain solutions better than the baseline within the first 50 iterations. SA-based requires more iterations to reach the baseline (within 100 iterations on bpic17 and 150 on wabo), due to its probabilistic acceptance of worse solutions, especially at the early phase (when the system temperature is high). However, observe that tree-based is quickly trapped in local optima on all datasets (the solution score remains unchanged after the first 100 iterations) while SA-based explores better-quality solutions. This is aligned with our expectations in terms of the design of the two methods.

Last but not least, we compared the efficiency of the two proposed methods based on the CPU time consumed to obtain the final solutions. Figure 3.7 illustrates the results. In general, the SA-based method requires more time to finish, particularly when solving problems that have a larger neighborhood to explore, i.e., bpic18 with 41 type-defining attributes and sepsis with 15. The larger neighborhood causes SA-based to stay at the same temperature for a longer period. Therefore, when the system temperature is high, the search has an increased possibility to explore worse solutions far from the initial solution. Note that in these experiments we chose to use empty initialization (i.e., starting a single execution context), which means those distant solutions are likely to correspond to larger sets of execution contexts — and our previous analysis of the time complexity of the algorithm (Section 3.3) has shown that the evaluation of such solutions is more costly. By comparison, the greedy tree-based method tends to stay within a relatively restricted part of the solution space — in this experiment, a subspace close to the initial, empty solution — and prevents itself from evaluating large-sized but worse solutions. Nevertheless, the longer time required by SA-based is a tradeoff for extensively exploring the solution space and increasing the possibility of obtaining better final solutions. As shown in (Figure 3.6), the final solutions produced by applying SA-based have higher quality scores compared to the tree-based ones.

3.4.5 Summary

Through the experiments above, we evaluated our approach to solving the problem of learning execution contexts. Our first goal was to test its feasibility. By comparing against the baseline execution contexts constructed for each dataset, we demonstrated that both proposed methods, i.e., tree-based and SA-based, are capable of utilizing information about the type-defining attributes in the log to learn compact and high-quality execution contexts. They performed well even when presented with complex problems with many type-defining attributes, i.e., log bpic18 with 41 and sepsis with 15.

Our second goal was to compare the two methods in detail. Our experiment results showed that the SA-based method outperformed tree-based by producing simpler execution contexts of better quality. An additional experiment showed that SA-based outputs could be further improved with more iterations allowed. Furthermore, we compared the efficiency of the two methods. We observed that tree-based was capable of obtaining relatively good-quality execution contexts usually within the first few iterations, while SA-based could produce better final outputs at the cost of time performance. These observations are aligned with our expectations of the two methods due to the different heuristics they employed.

3.5 Discussion

This chapter focuses on *execution context*, which is a fundamental notion in the *OrdinoR* framework that enables capturing the involvement of resource groups and their members in process execution. We introduced the problem of learning execution contexts from event logs and proposed to measure the quality of execution contexts based on how well they characterize resource specialization. We formulated the learning problem as utilizing event attributes to derive a set of so-called categorization rules that have maximized quality. Then, we proposed an approach based on decision tree learning and simulated annealing, respectively, to address the problem. We conducted experiments using five real-world event datasets. Based on our findings, we concluded that our approach is feasible for solving the problem. While the decision-tree-based and the simulated-annealing-based methods are both effective, the former runs more efficiently and the latter is capable of learning higher-quality execution contexts.

Our solution addresses the first task in the discovery of organizational models (Section 2.4). Given the essential role of execution contexts in the organizational models in the *OrdinoR* framework, this solution also contributes to better utilization of event log information for representing resource group involvement. In the meantime, note that having execution contexts is a prerequisite for the evaluation and analysis of organizational models. Hence, the solution introduced in this chapter is a key enabler of the overall organizational model mining approach (Figure 1.5) proposed in this book.

Beyond the scope of this research, learning execution contexts also contributes to other resource-oriented process mining topics focused on comparing resources and analyzing them along with other process dimensions, e.g., profiling resource behavior with regard to specific cases [68]. Since the learned execution contexts can be applied to select sub-logs to analyze process variants concerned with certain resources, our study also has potential contributions to the research on deriving process cube views in multidimensional process mining research [6, 19].

Our work has some limitations to be addressed in future work. From an input data perspective, further research is needed to investigate how event attributes with non-discrete values may be used directly as type-defining attributes, without having to be preprocessed. Those event attributes can be, for example, interdependent, continuous attributes that may not be discretized separately; or attributes that record key information used for decision-making during process execution but in the form of free-text. Dedicated solutions to the handling of such non-discrete attributes will enable application of the proposed approach on a broader range of event logs.

From an approach design perspective, it is worthwhile to consider impurity and dispersal as two separate optimization objectives instead of using a combined quality score. In that case, the proposed approach needs to return multiple sets of Pareto-optimal execution contexts — some with better impurity and others with better dispersal — and users can then select the desired one as the final solution. This way, we will be able to build a more "human-in-the-loop" approach that further incorporates user knowledge beyond what can be captured by input attribute specifications.

Another aspect to consider is to evaluate intermediate rules more efficiently in the iterative search. Note that this is currently done by directly computing impurity and dispersal. A more efficient way could be to minimize impurity and dispersal without incurring time complexity that is quadratic to the number of execution contexts. Devising such an efficient heuristic will contribute significantly to the application of the approach to large event logs.

For the simulated-annealing-based method, its configuration remains underexplored. With fine-tuned parameters and cooling schedules, this method could potentially be improved in terms of both output quality and efficiency. The issue can be investigated through multiple experiments on the same input dataset.

Chapter 4
Discovering Organizational Models

Abstract In the *OrdinoR* framework for organizational model mining, we outlined three tasks to be addressed in discovering organizational models from an event log (Section 2.4). Those are: (i) determining execution contexts based on the subimport log, (ii) determining resource grouping, i.e., groups of resources sharing similar behavior, and (iii) determining how to link execution contexts to resource groups to describe their involvement in process execution. Chapter 3 is devoted to solving the first task. This chapter introduces a systematic approach to discovering organizational models, covering all three tasks. We will look into the concrete challenges and discuss alternative methods for addressing them. We will also explain what and how user knowledge assists in configuring those methods.

4.1 A Three-Phased Discovery Approach

Figure 4.1 shows an overview of the approach to discovering organizational models from event logs. First, an event log with the standard attributes (*case*, *act*, *time*) and resource information (*res*) is used as input to determine a set of execution contexts. Using that, a resource-event log can then be derived and utilized for discovering resource groups, which includes characterizing the features of resources and clustering them into groups. Next, the discovered resource groups are "profiled" using information from the derived resource log to describe their group capabilities in process execution. As a result, an organizational model is constructed. Note that user knowledge plays an important role in configuring the methods in model discovery. Finally, discovered models can be evaluated and analyzed by applying the measures in the framework (Sections 2.5 and 2.6).

© The Author(s), under exclusive license to Springer Nature Switzerland AG 2026
R. J. Yang, *Mastering Organizational Dynamics Using Process Mining*, LNBIP 552, pp. 61–76, 2026.
https://doi.org/10.1007/978-3-031-93530-5_4

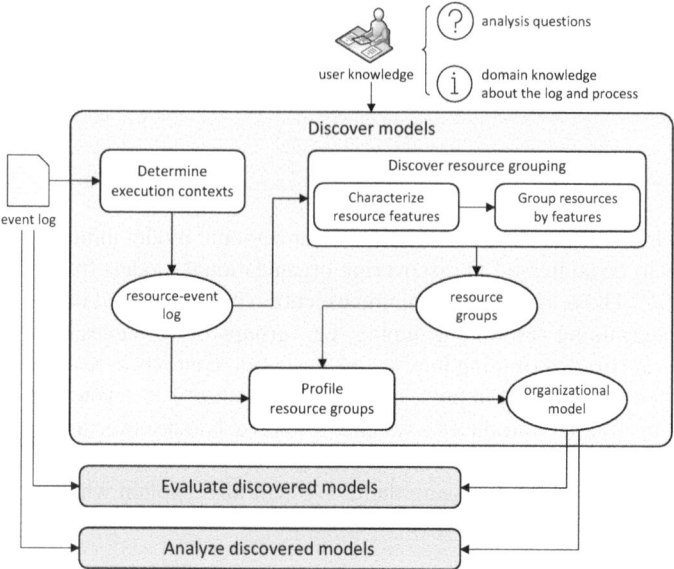

Fig. 4.1: An overview of the approach to the discovery of organizational models from event logs.

4.1.1 Determining Execution Contexts

There are two ways to determine execution contexts for a given event log — by directly specifying the types of case, activity, and time, or by applying a learning method to derive execution contexts from the log.

Direct type specification requires users to manually define both the names of types and how each type name corresponds to a category of cases, activity labels, or timestamps. A user, e.g., a process analyst or HR manager, may decide on type definitions using prior information about an event log and the recorded process and employees. Prior information can be questions that guide the current analysis and be based on domain knowledge about the process. For example, a process analyst is tasked to compare resources' performance by the types of customers they serve (analysis questions). In doing so, the process analyst is suggested by business experts that the process is designed to have a dedicated set of activities for handling high-end customers (domain knowledge). In this case, the process analyst may define case types by the customer types and define activity types by recognizing those specific activities.

When prior information is limited or unclear, some process mining techniques can be applied to help users decide on type specification. For example, trace clustering techniques (e.g., [78, 21]) are useful in finding coherent sets of cases, and behavioral patterns mining techniques (e.g., [84, 12]) can discover subsets of process activities representing frequent patterns in execution. These results may be utilized for directly

4.1 A Three-Phased Discovery Approach

deciding case types and activity types, or they can serve as knowledge in addition to the prior information.

Learning execution contexts from event logs provides an alternative means of determining execution contexts when direct type specification is not immediately applicable due to the lack of sufficiently concrete prior information. In Chapter 3, we formalized the learning execution contexts problem and introduced a solution that derives high-quality execution contexts from an event log, requiring minimal user domain knowledge as input. Compared to directly specifying types, learning execution contexts can exploit patterns embedded in event log data while still supporting the use of prior information about the categorization of cases, activity labels, and time.

4.1.2 Discovering Resource Grouping

Once a set of execution contexts is determined, an input event log is then transformed into a resource-event log (see Definitions 2.6 and 2.7), which is a sample describing resource behavior in process execution. Discovering resource grouping is concerned with how to use a resource-event log to identify groups of resources sharing similarities in their behavior.

To this end, the first step is to characterize resource features. Note that organizational models discovered from an event log should be descriptive of the reality as recorded in the log. Hence, we characterize resource features by a *resource-by-execution-context* matrix, which captures the variety and frequency of execution contexts in which resources performed work.

Given a resource-event log derived from an event log, a resource-by-execution-context matrix can be constructed using the number of occurrences of resource events. Table 4.1 shows a matrix that characterizes the features of the six resources in the example derived resource-event log (Table 2.2). Each row corresponds to a resource and each column corresponds to an execution context. A resource-by-execution-context matrix in practice usually has more rows and columns due to the larger numbers of events, event attributes, and resources recorded in real-life event logs.

Note that users may choose to configure a constructed resource-by-execution-context matrix to focus on dedicated resource features. This can be done by

- Context selection: Users may want to analyze specific execution contexts by the types of cases, activities, and times. Columns related to other execution contexts can thus be discarded; and
- Normalization: In some settings, users may want to exclude the workload difference across resources (e.g., considering full-time and part-time employees together [9, 77]). Or, they may want to omit the frequency difference across execution contexts, for example, there were more cases handled for normal customers compared to VIPs. To achieve these, matrix entries may be normalized

Table 4.1: An example resource-by-execution-context matrix related to the example resource-event log in Table 2.2.

resource	(normal, register, afternoon)	(normal, contact, afternoon)	(normal, check, morning)	(normal, decide, morning)	(VIP, register, morning)	(VIP, check, afternoon)	(VIP, decide afternoon)
Ann	0	1	0	0	0	0	0
Bob	0	0	0	0	1	0	0
John	0	0	1	1	0	0	0
Mary	0	0	0	0	0	1	1
Pete	3	0	0	0	0	0	0
Sue	0	0	1	1	0	0	0

by row sums (to exclude the difference across resources) and column sums (to omit the difference across execution contexts), respectively.

For instance, Table 4.2 shows the example resource-by-execution-context matrix configured for analyzing only the "VIP" cases and the one configured for excluding resource workload difference.

Table 4.2: Applying context selection to analyze only the "VIP" cases (left) and normalization by row sums to exclude workload difference (right) to the example resource-by-execution-context matrix in Table 4.1.

resource	(VIP, register, morning)	(VIP, check, afternoon)	(VIP, decide afternoon)
Ann	0	0	0
Bob	1	0	0
John	0	0	0
Mary	0	1	1
Pete	0	0	0
Sue	0	0	0

resource	*(execution contexts omitted for brevity)*						
Ann	0	100%	0	0	0	0	0
Bob	0	0	0	0	100%	0	0
John	0	0	50%	50%	0	0	0
Mary	0	0	0	0	0	50%	50%
Pete	100%	0	0	0	0	0	0
Sue	0	0	50%	50%	0	0	0

With a resource-by-execution-context matrix, we can address the task of identifying similar resources by applying established cluster analysis techniques in data mining. Agglomerative Hierarchical Clustering (AHC) [86] and KMeans [61, 14] are some of the classic algorithms, which generate disjoint clusters. More often than not, resource grouping in real-life organizations involves overlaps, i.e., resources may belong to more than one group. Hence, overlapping clustering (a.k.a. "soft clustering") techniques such as Model-based Overlapping Clustering (MOC) [15] and Gaussian Mixture Models (GMM) [42] can be applied to find potentially overlapping groups [88].

Most cluster analysis techniques require deciding the expected number of clusters. This is specified by users, indicating the number of potential resource groups

4.1 A Three-Phased Discovery Approach

they desire to discover from the log (e.g., a group number suggested by domain knowledge). Alternatively, users may have several candidates for the group number — in this case, silhouette score and cross-validation [70, 42] can be applied to help decide on the number of clusters.

4.1.3 Profiling Resource Groups

The final task is to profile each discovered resource group with a set of relevant execution contexts characterizing the group's capabilities in process execution.

We first consider a method, namely FullRecall, which accounts for all historical behavior by any member of a resource group. Given a derived resource-event log $RL(EL, CO)$, the set of execution contexts for profiling a group rg is

$$cap(rg) = \left\{ co \in CO \mid \exists_{r \in mem(rg)} (r, co) \in RL(EL, CO) \right\}. \tag{4.1}$$

Applying this definition, a resulting organizational model will capture all observed behavior recorded in the log and will thus achieve the best fitness. However, the use of FullRecall risks linking a resource group with an excessive number of execution contexts. This is because FullRecall considers every resource event related to any group member, even if that event may represent rare behavior.

Hence, we introduce another method OverallScore that ranks execution contexts according to how *frequent* and how *popular* they are with respect to the members of a resource group. If the process activities within an execution context were mostly taken by a specific group, or by the majority of members in a group, then this execution context is likely associated with the group.

OverallScore can be formalized as selecting execution contexts based on the weighted average of two model analysis measures, *group relative stake* (Definition 2.14) and *group coverage* (Definition 2.15), i.e.,

$$cap(rg) = \left\{ co \in CO \mid \omega_1 \cdot RelStake(rg, co) + \omega_2 \cdot Cov(rg, co) \geq \theta \right\}, \tag{4.2}$$

where θ is a threshold in the range $(0, 1)$, and ω_1, ω_2 are non-negative weights satisfying $\omega_1 + \omega_2 = 1$. These parameters can be set by users based on whether the main characteristic of group capabilities is reflected by relative stake (i.e., the group was the major participant) or coverage (i.e., most of the group members were involved).

Alternatively, users may perform a grid search to test multiple parameter settings and pick the one that produces a model with high quality based on fitness and precision (Section 2.5) — this model can then be selected as the discovery output. Compared to FullRecall, applying OverallScore links a resource group to only its most relevant execution contexts. This way, it leads to discovered models with relatively balanced fitness and precision, i.e., capturing most observed behavior in a log without allowing much excessive behavior.

Table 4.3 presents an example of profiling a group of three resources, using the two methods, respectively. Note that applying OverallScore excludes execution context "(normal, contact, afternoon)", which has low group coverage and thus an overall score lower than the given threshold.

Table 4.3: An example of profiling a resource group of three resources, applying FullRecall and OverallScore (setting weights $\omega_1 = \omega_2 = 0.5$ and threshold $\theta = 0.8$).

$mem(rg)$	$cap(rg)$ (applying FullRecall)	$cap(rg)$ (applying OverallScore)
Ann, John, Sue	(normal, contact, afternoon) (normal, check, morning) (normal, decide, morning)	(normal, check, morning) (normal, decide, morning)

4.2 Implementation

We developed an open-source software tool implementing the approach. It consists of (i) an extensible Python library[12] and (ii) a prototype web-based application, enabling users to perform organizational model mining tasks and visualize the outcomes. Figure 4.2 shows a screenshot of the prototype web application.

The tool has a modular design, following the proposed *OrdinoR* framework (Chapter 2). Several methods discussed in the previous sections have been implemented in the tool, as well as the proposed measures for evaluating and analyzing organizational models using event logs. The modular design of the tool also allows for extensions that introduce new methods and measures for organizational model mining in the future.

4.3 Evaluation

We conducted experiments on real-life event logs to demonstrate how to apply the proposed approach to discover organizational models using different alternative methods and how those methods compare. Furthermore, we show how to evaluate and analyze those discovered models using measures in the *OrdinoR* framework.

---

[12] The OrdinoR library: https://royjy.me/to/ordinor

4.3 Evaluation   67

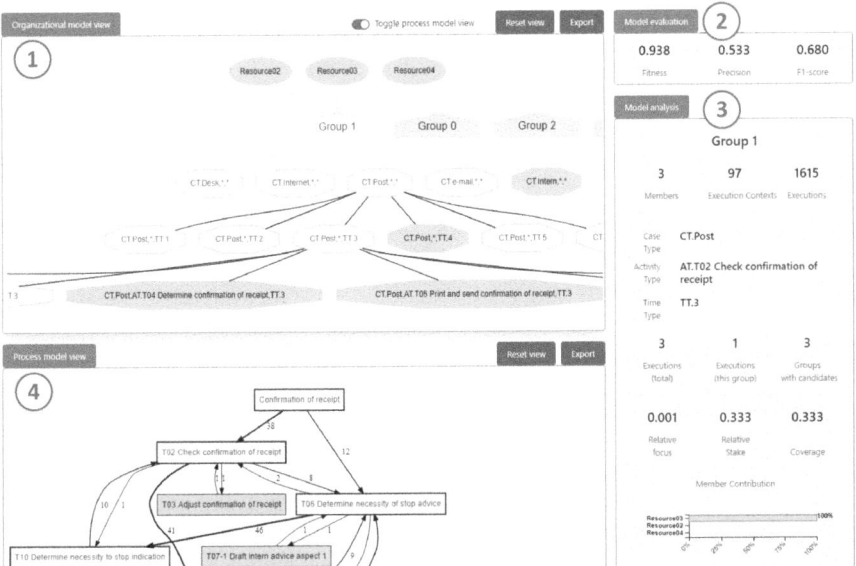

Fig. 4.2: An annotated screenshot of the software tool implementing the approach: (1) the visualization of a discovered organizational model; (2) the model's quality, measured by fitness, precision, and F1-score; (3) model analysis measures, along with some other descriptive statistics; (4) a Directly-Follows Graph representing the process model of the cases of the selected case type ("CT.Desk"), in which the red activities correspond to the activity types linked with the selected group ("Group 1").

4.3.1 Experiment Setup

Selecting datasets

The same collection of event logs introduced in Chapter 3 was used for the experiments here. Note that we utilized the preprocessed logs in order to incorporate the learning execution contexts outputs. For details on the experiment datasets, refer to Section 3.4.1.

Configuring the methods

Each of the three tasks in the model discovery approach can be addressed by alternative methods. In the experiments, we tested all combinations of these alternatives. Figure 4.3 depicts an overview of the experiment setup. Given an input event log, an organizational model is discovered by applying a combination of methods for each task and is then evaluated and analyzed.

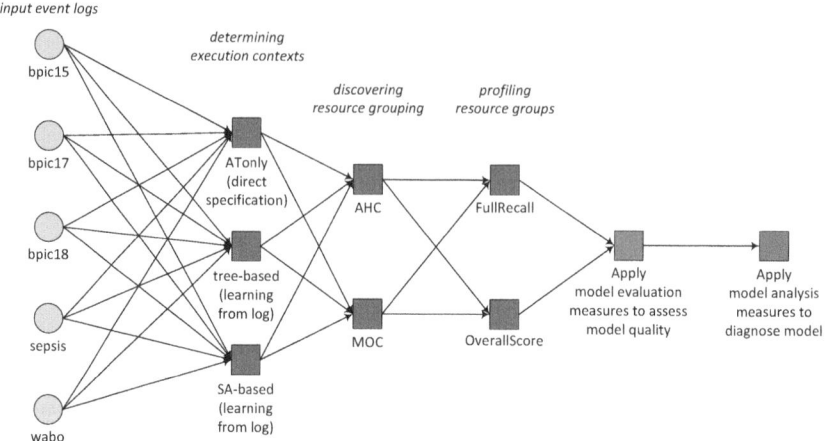

Fig. 4.3: An overview of the experiment setup: each path in the graph specifies a unique combination of methods for the three tasks. In total, there are 12 possible combinations of methods for discovering organizational models from an input event log.

Three alternative methods for determining execution contexts were tested. ATonly represents the method used by the majority of the organizational model mining literature, which considers only the process activities. This is equivalent to directly specifying the activity types by distinct activity names to construct execution contexts, with only a single case type and time type for all events. The other two are methods we developed for learning execution contexts, i.e., the tree-based method and the SA-based method (Section 3.3). To test them, we included the highest-quality execution contexts from the previous experiments (Section 3.4.4).

For discovering resource grouping, our experiments used the default resource-by-execution-context matrix, i.e., without considering any context selection or normalization. Two clustering techniques, Agglomerative Hierarchical Clustering (AHC) [86] and Model-based Overlapping Clustering (MOC) [15], were then applied to identify the resource groups. For their configuration, the Euclidean distance was selected as the proximity measure; the number of resource groups (clusters) was decided using cross-validation, in which the potential group number was tested between 2 and 10. Note that our experiments did not aim to investigate how an array of clustering techniques may perform on the discovery of resource grouping, hence the selected techniques were limited to the ones applied in the literature of organizational model mining [77, 88].

For profiling resource groups, method FullRecall requires no specific configuration. For OverallScore, we performed a grid search with a search step of 0.1 in the range $[0.1, 0.9]$ to determine the weights (ω_1, ω_2) and the threshold (θ).

4.3 Evaluation   69

Table 4.4: Discovered models with the best quality, used as baselines in the comparisons.

Log	Configuration	Model size		Model quality		
		#execution contexts	#resource groups	f.	p.	F1
bpic15	SA-based AHC OS	145	10	0.900	0.783	0.838
bpic17	SA-based AHC OS	2038	10	0.892	0.617	0.729
bpic18	SA-based AHC OS	62	10	0.978	0.938	0.957
sepsis	tree-based AHC OS	26	10	0.994	0.951	0.972
wabo	tree-based AHC OS	593	9	0.831	0.649	0.729

Configuration: OS = OverallScore
Model evaluation: f. = Fitness, p. = Precision, F1 = F1-score

4.3.2 Model Evaluation and Comparison

We discovered and evaluated a total of 60 organizational models (12 per event log). We compared models discovered from the same event log to investigate the impact of different model discovery methods on model quality. The baseline models used in the comparisons were the ones with the best quality, i.e., with the highest F1-score of model fitness and precision. Table 4.4 shows their size and quality.

We selected three subsets of organizational models for comparison against the baseline models. This selection corresponds to the three tasks in the model discovery approach. Note that each subset contains models discovered using different methods for *one* task, while for the other tasks the applied methods align with the corresponding baseline models. In the following, Table 4.5 reports the results of varying the methods for determining execution contexts; Table 4.6 reports the results of varying the methods for discovering resource grouping; and Table 4.7 reports the results of varying the methods for profiling resource groups. To aid the comparison, the results of the baseline models are underlined in these tables.

Determining execution contexts

(Table 4.5) Applying the tree-based or SA-based method produced models with the best quality and outperformed ATonly in general, especially in terms of model precision. This is because ATonly considers only the activity dimension, neglecting log information about different resource characteristics with regard to cases and times. Hence, the resultant models are less descriptive of the actual process execution and have lower quality. Models generated from applying tree-based and SA-based have comparable quality. Note that the SA-based models are simpler as they include fewer execution contexts — it may be more desirable to use these models, especially when sufficient time is allowed to tune and apply SA-based to determine execution con-

Table 4.5: Comparing models discovered by applying ATonly, tree-based, and SA-based to determine execution contexts, respectively.

Log	Configuration	Model size		Model quality		
		#execution contexts	#resource groups	f.	p.	F1
bpic15	ATonly AHC OS	495	10	0.857	0.601	0.706
bpic15	tree-based AHC OS	571	10	0.901	0.756	0.822
bpic15	SA-based AHC OS	145	10	0.900	0.783	0.838
bpic17	ATonly AHC OS	24	10	0.804	0.598	0.686
bpic17	tree-based AHC OS	3050	10	0.836	0.579	0.684
bpic17	SA-based AHC OS	2038	10	0.892	0.617	0.729
bpic18	ATonly AHC OS	18	10	0.950	0.924	0.937
bpic18	tree-based AHC OS	108	9	0.989	0.909	0.947
bpic18	SA-based AHC OS	62	10	0.978	0.938	0.957
sepsis	ATonly AHC OS	15	10	0.999	0.928	0.963
sepsis	tree-based AHC OS	26	10	0.994	0.951	0.972
sepsis	SA-based AHC OS	7	10	0.999	0.928	0.963
wabo	ATonly AHC OS	27	10	0.929	0.533	0.677
wabo	tree-based AHC OS	593	9	0.831	0.649	0.729
wabo	SA-based AHC OS	378	6	0.908	0.581	0.709

Configuration: OS = OverallScore
Model evaluation: f. = Fitness, p. = Precision, F1 = F1-score

texts. This conforms to our conclusions from the previous experiment (Section 3.4.4) comparing the two methods.

Discovering resource groups

(Table 4.6) The evaluation results show that AHC outperformed MOC, producing models with both higher fitness and precision. A possible reason is that the MOC-generated clusters are larger and have lower cohesion [82], i.e., data objects within the same cluster are more dissimilar, due to allowing overlaps between clusters. As shown in Figure 4.4, points representing clusters in the MOC models are generally located to the right and on top of those representing clusters in the AHC models, which indicates the larger cluster size and within-cluster distance. The poorer quality of the MOC clustering affects the subsequent task using OverallScore to profile the discovered resource groups (clusters), resulting in low-quality discovered models. We explain this below.

The low-quality MOC-generated clusters cause low model fitness. Recall from Section 4.1.3 that the OverallScore method considers an execution context as a resource group's capability if the execution context has sufficient relative group stake and group coverage. In the case of the MOC-generated clusters, their large size tends to lower group coverage; their low cohesion implies that resources are less likely to have contributed to the same execution contexts, which lowers relative stake. Conse-

4.3 Evaluation

Table 4.6: Comparing models discovered by applying AHC and MOC to discover resource grouping.

Log	Configuration	Model size		Model quality		
		#execution contexts	#resource groups	f.	p.	F1
bpic15	SA-based AHC OS	145	10	0.900	0.783	0.838
bpic15	SA-based MOC OS	145	10	0.784	0.773	0.778
bpic17	SA-based AHC OS	2038	10	0.892	0.617	0.729
bpic17	SA-based MOC OS	2038	9	0.793	0.630	0.702
bpic18	SA-based AHC OS	62	10	0.978	0.938	0.957
bpic18	SA-based MOC OS	62	4	0.764	0.820	0.791
sepsis	tree-based AHC OS	26	10	0.994	0.951	0.972
sepsis	tree-based MOC OS	26	10	0.977	0.929	0.952
wabo	tree-based AHC OS	593	9	0.831	0.649	0.729
wabo	tree-based MOC OS	593	10	0.754	0.611	0.675

Configuration: OS = OverallScore
Model evaluation: f. = Fitness, p. = Precision, F1 = F1-score

quently, fewer execution contexts will be linked with regard to the resource groups (clusters), which then leads to fewer events fitted by the discovered organizational models. In the meantime, the overlapped clusters generated by applying MOC allow resources to be members of multiple groups in the discovered models. This usually creates excessive candidate resources (see Definition 2.10) for events and causes lower model precision.

Profiling resource groups

(Table 4.7) Using FullRecall resulted in models with perfect fitness. But this method sacrifices precision, because resource groups are usually linked with a large number of irrelevant execution contexts. Consequently, FullRecall model may be too general — resources are allowed to carry out activities in excessive execution contexts, which is similar to the concept of "flower models" [7] in process model discovery, i.e., generic models that capture all observations in the data but are extremely imprecise. On the other hand, the baseline models (all resulted from applying OverallScore) have better precision while maintaining moderate fitness, since execution contexts were selectively linked to resource groups based on frequency and popularity.

Note that it is not necessary that a "flower-model" discovered by applying Full-Recall has low precision. In the case of log sepsis, both discovered models have decent precision values over 0.9. In fact, almost all the "flower-models" discovered from that log applying FullRecall have satisfactory precision, except one that was generated from applying tree-based-MOC-FullRecall. Next, we utilized the model analysis measures to further investigate this exception.

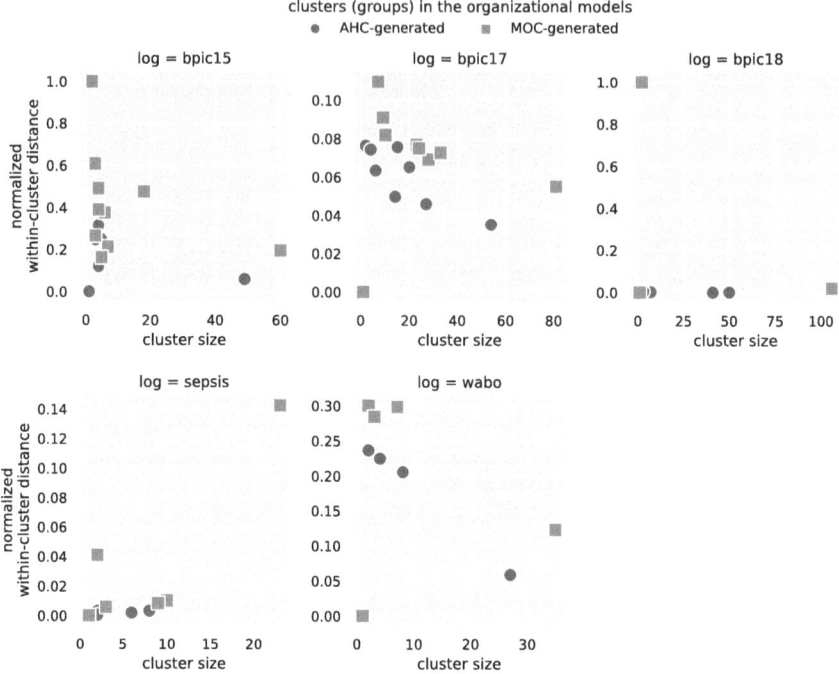

Fig. 4.4: Size and cohesion (measured by normalized within-cluster distance) of the clusters in the models discovered by applying AHC and MOC (Table 4.6). Note that higher within-cluster distance (y-axis) implies lower cohesion.

4.3.3 Model Diagnosis

The model to be diagnosed was discovered from log sepsis using tree-based-MOC-FullRecall. It has low quality due to poor precision (precision = 0.104, F1-score = 0.188). All other 11 models discovered from log sepsis have high quality, with an average precision of 0.929 and an F1-score of 0.960, leaving the selected model as an "outlier". We applied the model analysis measures to reveal the cause.

The perfect fitness and poor precision of the outlier model imply that some resource groups and execution contexts were inappropriately linked during discovery, causing certain events in the log to have excessive candidate resources (Definition 2.10). To identify such groups and execution contexts, we applied the *group relative stake* (Definition 2.14), *group coverage* (Definition 2.15), and *group member contribution* (Definition 2.16) measures. Group relative stake can reveal the amount of a group's contribution to an execution context. Group coverage can show the proportion of group members involved in an execution context, and group member contribution can then be used to reveal those involved members. Table 4.8 presents the average values of group relative stake and group coverage for each group in the

4.3 Evaluation

Table 4.7: Comparing models discovered by applying FullRecall and OverallScore to profile resource groups.

Log	Configuration	Model size		Model quality		
		#execution contexts	#resource groups	f.	p.	F1
bpic15	SA-based AHC FR	145	10	1.000	0.259	0.411
bpic15	SA-based AHC OS	145	10	0.900	0.783	0.838
bpic17	SA-based AHC FR	2038	10	1.000	0.225	0.367
bpic17	SA-based AHC OS	2038	10	0.892	0.617	0.729
bpic18	SA-based AHC FR	62	10	1.000	0.226	0.369
bpic18	SA-based AHC OS	62	10	0.978	0.938	0.957
sepsis	tree-based AHC FR	26	10	1.000	0.928	0.962
sepsis	tree-based AHC OS	26	10	0.994	0.951	0.972
wabo	tree-based AHC FR	593	9	1.000	0.228	0.372
wabo	tree-based AHC OS	593	9	0.831	0.649	0.729

Configuration: FR = FullRecall, OS = OverallScore
Model evaluation: f. = Fitness, p. = Precision, F1 = F1-score

outlier model. In addition, to compare the groups, we calculated their rankings based on those average values.

Table 4.8: Average group relative stake and group coverage of the resource groups in the outlier model (discovered from sepsis using tree-based-MOC-FullRecall). The two groups in bold text ("Group 1" and "Group 3") were pinpointed by the model diagnosis for detailed analysis. Note that the resource group names were randomly assigned by the applied clustering technique.

resource group	#group capabilities	#group members	average group relative stake	rank	average group coverage	rank	sum of ranks
Group 1	26	23	0.972	7	0.114	1	8
Group 2	6	1	0.948	6	1.000	8	14
Group 3	6	2	0.116	1	0.500	3	**4**
Group 4	6	1	1.000	9	1.000	8	17
Group 5	3	1	1.000	9	1.000	8	17
Group 6	3	10	0.576	5	0.733	4	9
Group 7	2	1	0.194	2	1.000	8	10
Group 8	5	9	0.490	4	0.489	2	6
Group 9	3	3	0.343	3	0.889	5	8
Group 10	5	1	1.000	9	1.000	8	17

From the table, it can be observed that "Group 1" is problematic. It includes 23 member resources and was profiled with all 26 execution contexts as group capabilities. The reason is likely that MOC generated an abnormally large cluster (including 23 out of the total 25 resources) — which is then profiled with all execution contexts by FullRecall indiscriminately, causing high group relative stake but low group coverage.

We confirmed it by examining the distribution of group coverage of all the execution contexts with regard to "Group 1". As shown in Figure 4.5, most of the execution contexts profiled as this group's capabilities have group coverage lower than 0.2, i.e., only a small proportion of group members were involved in these execution contexts. Consequently, all the members are considered candidate resources for all the execution contexts, which explains the poor precision of the model.

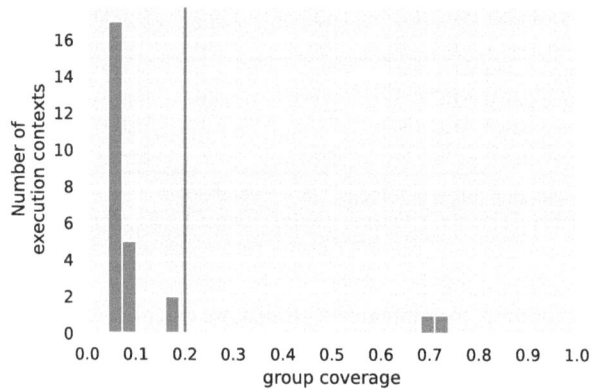

Fig. 4.5: Distribution of group coverage values of all execution contexts with regard to "Group 1". Notice that most of the execution contexts have group coverage lower than 0.2.

But, not all problematic parts stand out like "Group 1". We also investigated "Group 3" consisting of two resources as another example. This group has the lowest average group relative stake and group coverage based on the rankings. Table 4.9 shows the six execution contexts as the group's capabilities and their group relative stake, coverage, and member contribution. All the execution contexts have coverage of just 0.5, indicating that the outlier model over-generalizes execution contexts specific to a single group member as capabilities of both members. We can see that the two resources, i.e., "F" and "L", have different specializations in terms of activity types — "F" performed the activity type "AT.3" (transfer to normal care), while "L" performed the others (giving infusions, ER Registration, and ER Sepsis Triage). This suggests that "F" and "L" could have been placed in two groups. Instead, the outlier model included them in the same group.

Based on the diagnosis findings above, we created an organizational model that improves the outlier model by (i) discarding the abnormal "Group 1" and (ii) splitting the group of "F" and "L" into two singletons, each linked with the execution contexts specific to a resource as group capabilities. Compared to the original model, the improved one has nearly perfect fitness (0.993) and a much improved precision

4.3 Evaluation

Table 4.9: Capabilities of "Group 3" in the outlier model, measured by group relative stake, group coverage, and group member contribution per each resource in the group.

execution context	group relative stake	group coverage	member (resource id)	group member contribution
(CT.0, AT.3, TT.0)	0.208	0.500	F	100%
(CT.0, AT.3, TT.1)	0.180	0.500	F	100%
(CT.0, AT.0, TT.0)	0.102	0.500	L	100%
(CT.0, AT.0, TT.1)	0.050	0.500	L	100%
(CT.0, AT.7, TT.0)	0.099	0.500	L	100%
(CT.0, AT.7, TT.1)	0.059	0.500	L	100%

- CT.0 is case type of which cases are with "no ordered diagnostics for liquor".
- AT.0 is an activity type of "giving infusions of liquid and antibiotics";
- AT.3 is an activity type of "admission or transfer to normal care";
- AT.7 is an activity type for "ER Registration" and "ER Sepsis Triage".
- TT.0 is a time type of the calendar month January;
- TT.1 corresponds to all other months.

(0.924). This result supports our diagnoses about "Group 1" and "Group 3" being the key problems in the outlier model.

4.3.4 Summary

Through the experiments, we demonstrated that the proposed approach is capable of discovering organizational models with satisfactory quality. As Table 4.4 displays, the best-quality discovered models achieved F1-scores of at least 0.7, with two models having F1-scores over 0.9. We also showed that several alternative methods can be applied to address the three tasks in model discovery and compared them based on the fitness and precision of the resultant models. The comparison results provide insights into the selection of techniques, which can benefit future applications of the discovery approach on other event logs:

1. the tree-based and SA-based learning execution contexts methods proposed in Chapter 3 are effective for the task of determining execution contexts, and exploiting multidimensional process information leads to discovering better quality organizational models;
2. AHC outperforms MOC when applied to discover resource grouping, due to the fact that MOC — as an overlapping clustering technique — may risk generating clusters that are over-sized and less cohesive and hence cause a decrease in model fitness and precision;
3. OverallScore is an effective method, compared to FullRecall, for profiling resource groups, as it generates organizational models with balanced fitness and precision.

The model diagnosis showed how to apply the model analysis measures to uncover the problems behind models with unsatisfactory quality. Using those measures, we were able to pinpoint problems inside the poor-quality model discovered from log sepsis. We revealed two contributing factors to the model's low precision: (i) resource group being too generic ("Group 1"), and (ii) resource group consisting of dissimilar members ("Group 3"). We also demonstrated that the model analysis outcomes are useful in "repairing" a problematic organizational model having those issues.

4.4 Discussion

This chapter introduces an end-to-end approach to the discovery of organizational models from event logs, addressing the three tasks outlined in the *OrdinoR* framework. Execution contexts can be determined by directly specifying the types based on prior information about event logs and processes; or they can be learned from event logs by using the approach proposed in Chapter 3. Resource grouping is identified by first constructing a resource-by-execution-context matrix to characterize resource features and then applying conventional clustering techniques. Lastly, execution contexts are linked with resource groups as their capabilities, based on either recalling all execution contexts that the group members were involved in or selecting the ones that are sufficiently relevant to the group members. We evaluated our approach through experiments on the same collection of real-world event logs used in Chapter 3. In the first part of the experiments, we validated the effectiveness of the proposed approach and analyzed how applying different techniques impacts the discovered models' quality. In the second part of the experiments, we demonstrated the usefulness of the model analysis measures in enabling a detailed diagnosis of low-quality organizational models.

The proposed approach contributes a realization of the *OrdinoR* framework and shows the application of various techniques. It offers a solution for discovering organizational models from event logs. In the meantime, the experiments highlight the value of model evaluation and model analysis in the framework. First, the experiments can be viewed as a validation of the proposed model evaluation and analysis measures. Second, the experiment results showed that using fitness and precision can provide a basis for objectively and independently assessing model quality, while the use of model analysis measures contributes a way to explain the model evaluation results. Together, they fill the gaps identified in the state-of-the-art of organizational model mining.

Future work may extend the model discovery approach by introducing a wider variety of novel techniques to improve the quality of discovered models. This calls for a comprehensive benchmark of different techniques and their various configurations. In doing so, it is worthwhile creating artificial event logs to test the approach under scenarios that are likely to happen, but not captured in existing, real-world datasets. Also, it will be useful to explore how the characteristics of event logs, their processes, and the organizations may inform the selection and configuration of the techniques employed in the approach.

Chapter 5
Applying Organizational Models to Workforce Analytics

Abstract Event logs are useful data sources for deriving knowledge about the organizational grouping of human resources in the context of business process execution. In the previous chapters, we concentrated on discovering organizational models that effectively characterize resources, their grouping, and their involvement along multiple process dimensions. We proposed approaches to automatically constructing such models with minimum data requirements and evaluating discovered models to ensure that they capture the organizational information stored in event logs completely and exactly (i.e., achieving good model fitness and precision). In this chapter, we will focus on the application of organizational models to support workforce analytics concerned with employee groups. This is built upon the organizational model analysis in the *OrdinoR* framework: extending an organizational model with the temporal information about events and cases in an event log, so that the behavior of resource groups and their members can be examined. In Chapter 4, we focused on using this idea for diagnosing low-quality discovered models, that is, to locate issues that cause a model to deviate from the subimport event log (Section 4.3.3). Here, we enhance the idea for a different purpose — we aim at utilizing event logs to create "profiles" of resource groups to quantitatively characterize how they work in business process execution, from various aspects and across different periods. Specifically, we will look into what aspects can be measured as the work profiles of resource groups, and will discuss how these measures can be analyzed to provide insights into managing resource groups.

5.1 Preliminaries

To explain the profiling of resource groups from various aspects, we first introduce the following auxiliary notation for organizing events in a log. $\mathcal{T}$ is the universe of timestamps, and $[t_1, t_2)$ denotes a half-open time interval with $t_1, t_2 \in \mathcal{T}$ and $t_1 < t_2$. Let $EL = (E, Att, \pi)$ be an event log and let $OM = (RG, mem, cap)$ be an organizational model with a set of pre-defined execution contexts $CO = rng(cap)$, then

- given an execution context $co \in CO$, $[E]_{co}$ denotes the set of events in EL corresponding to co (Definition 2.4);
- given a resource group $rg \in RG$,

$$[E]_{rg} = \{\, e \in E \mid \pi_{res}(e) \in mem(rg) \,\}$$

denotes the set of events in EL originated by resources in rg;
- given a time interval $[t_1, t_2)$,

$$[E]_{t_1,t_2} = \{\, e \in E \mid \pi_{time}(e) \in [t_1, t_2) \,\}$$

denotes the set of events in EL originated between t_1 (inclusive) and t_2.

We also define some auxiliary notation for organizing cases in an event log.

- Given an execution context $co = (ct, at, tt) \in CO$,

$$[EL]_{ct}^{case} = \{\, c \in rng(\pi_{case}) \mid c \in \varphi_{case}(ct) \,\}$$

denotes the set of cases in EL having case type ct;
- given a resource group $rg \in RG$,

$$[EL]_{rg}^{case} = \{\, c \in rng(\pi_{case}) \mid \exists_{e \in E} [\pi_{case}(e) = c \land \pi_{res}(e) \in mem(rg)] \,\}$$

denotes the set of cases in EL that involved members of rg;
- given a time interval $[t_1, t_2)$,

$$[EL]_{t_1,t_2}^{case} = \left\{\, c \in rng(\pi_{case}) \;\middle|\; \exists_{e \in [E]_{t_1,t_2}} [\pi_{case}(e) = c] \,\right\}$$

denotes the set of cases in EL with at least one event occurrence between t_1 (inclusive) and t_2;
- given a time interval $[t_1, t_2)$,

$$[EL]_{t_1,t_2,\text{complete}}^{case} = \{\, c \in [EL]_{t_1,t_2}^{case} \mid \nexists_{e' \in E} [\pi_{case}(e') = c \land \pi_{time}(e') \geq t_2] \,\}$$

denotes the set of cases in EL completed between t_1 (inclusive) and t_2.

In addition, for a case in the log $c \in rng(\pi_{case})$ that is completed, we use $\tau(c)$ to denote the case cycle time, i.e., the duration from the first event to the last event. Formally, let $e_{start}^c \in E$ such that $\nexists_{e' \in E} [\pi_{case}(e') = c \land \pi_{time}(e') < \pi_{time}(e_{start}^c)]$, and $e_{end}^c \in E$ such that $\nexists_{e' \in E} [\pi_{case}(e') = c \land \pi_{time}(e') > \pi_{time}(e_{end}^c)]$, then we have

$$\tau(c) = \pi_{time}(e_{end}^c) - \pi_{time}(e_{start}^c).$$

5.2 Resource Group Work Profiles

Drawing on the theoretical and conceptual background in the prior section, this section presents the notion of *work profile of resource groups*, inspired by research on mining individual resource behavior [68, 49, 80]. A work profile of a resource group can be defined as a collection of *indicators* used to measure different *aspects* of that group of resources, in terms of their interaction with the relevant work in process execution. As with any indicators related to performance, the measurement of indicators is temporally aware, i.e., considering a time interval between t_1 and t_2, in which the respective performance of a group is measured [27]. By specifying the relevant interval, work profiles can reflect the fact that the performance of resource groups is often dynamic due to resources having shifts and turnover.

Definition 5.1 (Work Profile of a Resource Group) Let *RG* be a set of resource group identifiers, $\mathcal{T}$ the universe of timestamps, and $[t_1, t_2)$ a half-open time interval with $t_1, t_2 \in \mathcal{T}$ and $t_1 < t_2$. Let $\mathcal{I}$ be a set of names for possible indicators. Given a resource group $rg \in RG$, $WP = (rg, t_1, t_2, \mathcal{I}, \lambda)$ is a work profile for the resource group during time period $[t_1, t_2)$, where $\lambda : \mathcal{I} \to \mathbb{R}$ specifies the quantified measures of the indicators.

The definition provides a general representation of indicators measuring different aspects of a resource group over a specific time frame.

5.2.1 Work Profile Indicators

By reviewing the management literature, we identified a number of studies on human resource performance measurement [20, 23, 27, 41, 45] that can inform the proposal of a resource group's work profile useful for workforce analytics. The indicators correspond to the input-throughput-output view on processes [28]: *Performance* regarding input-output can be measured with indicators related to productivity and efficiency. Whether a specific output is achieved is referred to as *goal achievement*. Finally, the throughput is reflected by the summation of employee *workload* in a group. As a result, we present a collection of three general aspects and the associated indicators, focusing on a resource group in its entirety.

Workload [23]: *What and how much work is a resource group involved in?* This can be measured by

- allocation, the overall amount of work allocated to the group;
- assignment, the amount of the group's workload assigned to specific work;
- relative focus, the proportion of the group's workload assigned to specific work; and
- relative stake, the amount of contribution by the group to specific work.

Performance [20, 27, 41, 45]: *How does a group perform?* This can be measured by

- amount-related productivity, the amount of work completed by the group;

- time-related productivity, the time required by the group to complete the work; and
- efficiency, the amount of satisfactory work produced by the group.

Goal achievement [20, 41]: *To what extent does a group adhere to goals?* This can be measured by effectiveness, i.e., the proportion of established goals accomplished by the group.

In this research, we also consider how resource groups interact with work in terms of their involvement in business process execution captured by event logs. This is reflected in the following three aspects and their indicators, which measure how group members interact with relevant work in a process and with each other.

Participation [20, 27]: *How do group members commit to work?* This can be measured by attendance, the number or proportion of group members committing to work.

Distribution [20]: *How is work distributed over group members?* This can be measured by

- member load, the amount of work allocated to individual group members, and
- member contribution, the amount of specific work contributed by an individual group member.

Collaboration [27]: *How is the collaboration among group members?* This can be measured as cooperation, i.e., the extent of collaboration between group members.

The above collection of six aspects and associated indicators can be used to form the template of a group's work profile for group-oriented analysis. Note that the term "work" here refers to either the activities or cases in business process execution.

5.2.2 Extracting and Analyzing Work Profiles

We introduce an approach to extracting and analyzing work profiles of resource groups using event logs. Figure 5.1 depicts an overview of the proposed approach consisting of two phases.

Extraction of Work Profiles

The approach starts with determining resource groups and execution contexts. The first input is an event log, which should satisfy the minimum requirements by recording at least the standard attributes and the resource identifier (Definition 2.2). An organizational model is required as the second input, which can be obtained through model discovery from event logs (Chapter 4). Alternatively, domain knowledge that informs execution contexts and resource groups may be used as input when (i) there exists clear information on case types, activity types, time types, and resource grouping that users wish to use for analyses; or (ii) discovered organizational models do not have satisfactory quality. Note that the types should be determined through the

5.2 Resource Group Work Profiles

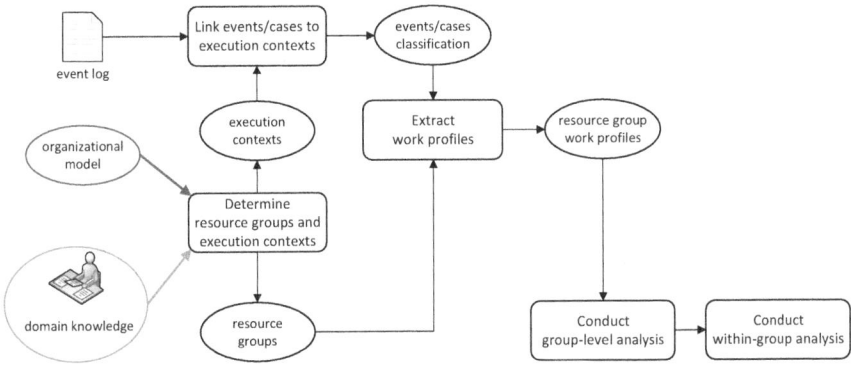

Fig. 5.1: An overview of the approach to extracting and analyzing resource group work profiles. Note that an organizational model or domain knowledge can be used alternatively as input.

direct type specification (Section 4.1.1), i.e., defining type names and their correspondence to the categorization of cases, activity labels, and timestamps. Similarly, any domain knowledge about resource groups should correspond to the grouping of resource identifiers in the log. The above requirements on input domain knowledge are essential to ensuring that the knowledge can be used as an alternative to an input organizational model.

The next step is linking events and cases in the input log to the execution contexts, which is straightforward given the mapping between types and event attributes. The output is the classification of events and cases, i.e., for any execution context, we can retrieve a unique set of events and a set of related case identifiers from the event log.

Then, the indicators of work profiles of resource groups can be calculated. We formally describe the pre-defined work profile indicators (Section 5.2.1) that can be directly extracted given an event log with essential information recorded. Given an event log $EL = (E, Att, \pi)$, a set of resource groups RG and their members $mem: RG \rightarrow \mathcal{P}(\mathcal{R})$, and a set of execution contexts CO, the pre-defined work profile indicators can be measured for a resource group $rg \in RG$ and a time interval $[t_1, t_2)$ as follows.

Workload: The indicators of resource group workload capture the amount of different types of work carried out by a resource group. With respect to an event log, the amount of work can be quantified by considering either the number of activities (which can be inferred from the number of unique events) or the number of cases (which can be inferred from the number of unique case identifiers).

- allocation is measured by the total number of activities conducted by a group, $|[E]_{rg} \cap [E]_{t_1,t_2}|$, or by the total number of cases involving the group, $|[EL]_{rg}^{case} \cap [EL]_{t_1,t_2}^{case}|$;
- assignment is measured by the number of activities conducted by a group that are specific to some execution context (co), $|[E]_{rg} \cap [E]_{t_1,t_2} \cap [E]_{co}|$, or by

the number of cases involving a group that are specific to some case type (ct), $\left|[EL]_{rg}^{case} \cap [EL]_{t_1,t_2}^{case} \cap [EL]_{ct}^{case}\right|$;

- relative focus measures the assignment of specific activities to a group in proportion to the group's allocation. We proposed this as a model analysis measure (Definition 2.13). Here, we extend it to consider a selected time interval, i.e., counting only events in $[E]_{t_1,t_2}$. To measure relative focus based on cases, one can calculate the proportion of assignment to allocation by case number, that is, $\left|[EL]_{rg}^{case} \cap [EL]_{t_1,t_2}^{case} \cap [EL]_{ct}^{case}\right| / \left|[EL]_{rg}^{case} \cap [EL]_{t_1,t_2}^{case}\right|$.
- relative stake measures the assignment of specific activities to a group in proportion to the total execution of those activities captured by the event log. We proposed this as a model analysis measure (Definition 2.14). Similar to relative focus, here we count only events and cases in $[E]_{t_1,t_2}$: relative stake based on events is $\left|[E]_{rg} \cap [E]_{t_1,t_2} \cap [E]_{co}\right| / \left|[E]_{t_1,t_2} \cap [E]_{co}\right|$; relative stake based on cases is $\left|[EL]_{rg}^{case} \cap [EL]_{t_1,t_2}^{case} \cap [EL]_{ct}^{case}\right| / \left|[EL]_{t_1,t_2}^{case} \cap [EL]_{ct}^{case}\right|$.

Performance: The indicators of group performance can be quantified by considering cases completed in a given time interval. Let $[EL]_{rg,t_1,t_2,\text{complete}}^{case} = [EL]_{rg}^{case} \cap [EL]_{t_1,t_2,\text{complete}}^{case}$ be cases completed in interval $[t_1, t_2]$ by a group, then

- amount-related productivity is measured by the total number of completed cases by the group, $\left|[EL]_{rg,t_1,t_2,\text{complete}}^{case}\right|$;
- time-related productivity is measured by the average time taken by the group to complete those cases,

$$\left(\sum_{c \in [EL]_{rg,t_1,t_2,\text{complete}}^{case}} \tau(c)\right) / \left|[EL]_{rg,t_1,t_2,\text{complete}}^{case}\right|;$$

- efficiency extends amount-related productivity by including some pre-defined normative criteria. For example, an analyst can specify that only cases completed within 10 days are considered "satisfactory", and therefore efficiency will be calculated based on the number of satisfactory cases by the group only.

Goal achievement: The effectiveness indicator measuring the *goal achievement* of a resource group is quantified based on other aspects and their indicators. For example, given two goals established in terms of the maximum amount of allocation (measuring workload) and the minimum level of efficiency (measuring performance), the effectiveness of a group can be measured by considering whether the group accomplishes these goals, respectively.

Participation: The indicator attendance can be quantified by considering the occurrences of group members carrying out activities or cases. Note that this should be considered a rough estimate, since an event log may not accurately capture the time when employees *started* working on a process. Let $r \in mem(rg)$ denote a member of a resource group rg, then

- attendance is measured by the number of member resources in a group who originated at least one event, $\left|\left\{r \in mem(rg) \mid \exists_{e \in [E]_{t_1,t_2}} \pi_{res}(e) = r\right\}\right|$.

5.2 Resource Group Work Profiles

Distribution: The indicators for distribution are defined over group members by calculating the portion of workload of the group. Again, consider $r \in mem(rg)$ a member of a resource group rg,

- member load is measured by the number of activities performed by a resource, $|\{e \in [E]_{rg} \cap [E]_{t_1,t_2} \mid \pi_{res}(e) = r \wedge r \in mem(rg)\}|$. Clearly, the sum of member load across all members of a group should be equal to the activity allocation to the group;
- member assignment is measured in a similar way, but considering only specific activities that are part of an execution context co, that is, $|\{e \in [E]_{rg} \cap [E]_{t_1,t_2} \cap [E]_{co} \mid \pi_{res}(e) = r \wedge r \in mem(rg)\}|$.

Collaboration: Quantifying the extent of collaboration among employees using event logs can be challenging, since (i) event logs usually do not capture the communication between employees and (ii) the way collaboration happens in different processes and organizations may vary. We consider a possible way to *estimating* cooperation. One can use the event log to construct the handover-of-work network [9] of group members, which reflects the frequency of work transfer between resources in process execution. Then, the cooperation of the group can be measured by the density of the handover-of-work network — a larger density indicates that there is more work transfer and hence a higher level of cooperation within the group.

Analysis of Work Profiles using Visual Analytics

Building on work profiles extracted from event logs, different data analytics techniques can be applied to discover patterns from the measurement of indicators. In our approach, we discuss the use of visual analytics as an intuitive and proven means [47] for analyzing work profiles. Following the definition of work profiles and the relevant aspects and indicators, we consider the requirements below for visually analyzing work profiles.

- Users should be able to interactively extract work profiles related to different time intervals in an event log and at different granularity (e.g., daily, monthly), and therefore track the changes of work profiles over time.
- Users should be able to have an integrated view of interrelated indicators (e.g., allocation and assignments) to derive findings on interactions between different aspects or process dimensions.
- Users should be able to compare indicators measured among different groups at different times.
- Users should be able to correlate indicators for group-level analysis with those for within-group analysis to obtain a holistic view of groups' work behavior.

Based on these requirements and guided by the general principles of visual analytics [52], we developed a design composed of several types of charts combined with interactive filters. The design aims to provide an integrated and purposeful visualization of multiple aspects of a resource group's work profiles. The following is included.

- A *stacked area chart* and a *line chart* are chosen for analyzing workload and performance, given their advantages in capturing indicator values as time series and showing the evolution patterns. For these two charts, interactive filters are embedded to allow users to explore the workload and performance indicators at different times and different levels of granularity.
- A *heatmap* is used for supporting the analysis of workload and distribution with regard to different case, activity, and time types, for its usefulness in simultaneously presenting values related to two-dimensional data attributes.
- A *stacked bar chart* is used for presenting intuitively the attendance of group members with respect to group size.

By connecting different charts using the same set of interactive filters, users are provided with an integrated view of work profiles of resource groups in a selected time interval of interest.

We implemented a prototype built upon Vega-Lite [72]. Figure 5.2 and Figure 5.3 illustrate the prototype's interactive visualization interface. The tool is publicly available online[13].

---

[13] Link to the prototype tool: https://royjy.me/to/gwp-demo

5.2 Resource Group Work Profiles

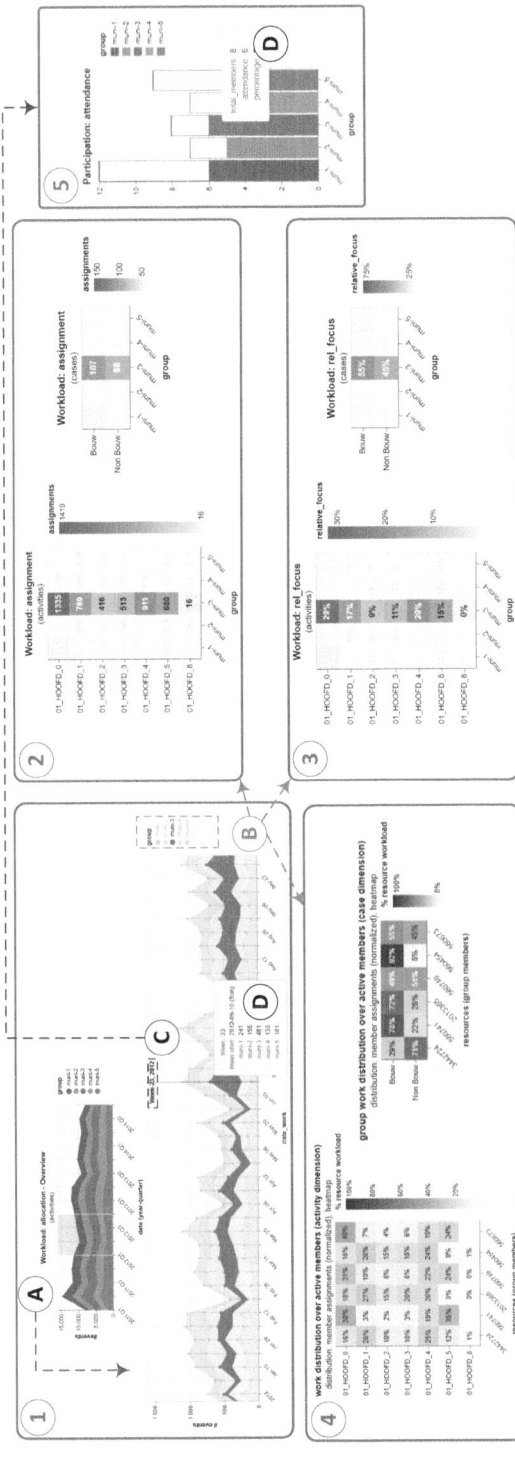

Fig. 5.2: Annotated screenshots of the prototype's interactive interface for analyzing work profiles regarding workload, participation, and distribution. The numbers mark different views: (1) workload by allocation; (2) workload by assignment measuring either activities or cases; (3) workload by relative focus measuring either activities or cases; (4) distribution by member assignment; (5) participation by attendance. The views respond to user interactions simultaneously: (A) selecting a time interval and zoom-in; (B) highlighting specific groups; (C) focusing on a specific time period (week); and (D) showing specific numbers via a tooltip. Note that these screenshots are for demonstrating the use of various charts and their integration, and the text within the screenshots is not of primary relevance.

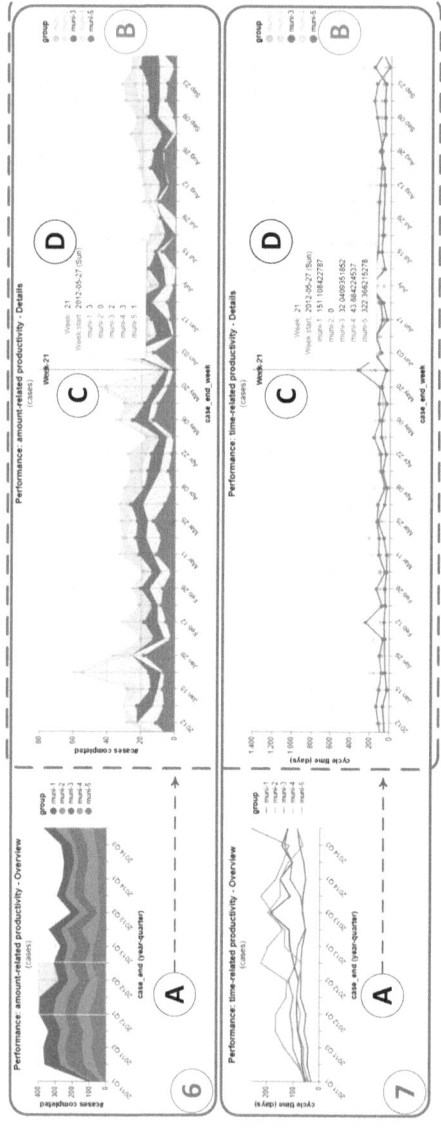

Fig. 5.3: Annotated screenshots of the prototype's interface for analyzing work profiles regarding performance. Views of (6) amount-related productivity and (7) time-related productivity respond simultaneously to user interactions (A–D). Note that these screenshots are for demonstrating the use of various charts and their integration, and the text within the screenshots is not of primary relevance.

The design shows a possible way of applying visual analytics to analyze work profiles. While the aspects and indicators of a work profile may be further extended, other visualization techniques can be applied accordingly.

Next, we will demonstrate how the proposed approach can be applied to conduct a resource-group-oriented analysis. We will use a real-life event log dataset, bpic15, which records a building permit application process performed in five Dutch municipalities. For details on this dataset, refer to Section 3.4.1.

5.3 Case Study: One Process, Five Municipalities

We used the bpic15 dataset to conduct a case study and tested our approach. The dataset captures how an identical building permit handling process was performed in five different municipalities in an approximate four-year period. The process owners raised a few business questions, aiming to better understand the differences between the municipalities and their impact on performance. Some of these questions are as follows.

1. Where are differences in throughput times between the municipalities and how can these be explained?
2. What are the roles of the people involved in the various stages of the process and how do these roles differ across municipalities?

bpic15 serves as a representative example for scenarios where multiple resource groups perform similar work and the managers wish to compare them and derive implications for future improvements. Given this context, we considered each municipality as a resource group in our evaluation and applied the approach to extract and analyze their work profiles.

We preprocessed the original data to facilitate the analyses, guided by the business questions. To ensure a fair comparison across the five groups, we set the scope of analysis to the main subprocess of handling cases between year 2011 to 2014. To this end, we first filter events recording the main subprocess (with "01_HOOFD" as the "subprocess" attribute value). Then, we keep only cases that started no earlier than 2011-01-01 and were completed no later than 2014-12-31. Also, we discarded cases that have invalid cycle time recorded, i.e., the duration between the first and the last event should be greater than 0. Lastly, we discarded a few cases that were handled by employees from more than one municipality — so that the case cycle time can indicate the performance of each individual group.

Clearly, the given business questions are concerned with the existing grouping of resources, i.e., the five municipalities. Hence, in this case study, we chose not to use discovered organizational models, but instead manually determine the execution contexts through direct type specification. For case types, we distinguished between cases that were related to a construction permit and those unrelated. This can be determined by the value of a derived case attribute, "case:parts Bouw". For activity types, we considered the different phases in the process, extracted from the unique values of a derived event attribute "phase". For time types, we used the seven days of the week, i.e., Monday to Sunday.

As a result, the preprocessed dataset used for analyses contains a total of 167691 events from 4792 cases involving 61 resources in the five resource groups (municipalities). The defined execution contexts consist of two case types, nine activity types, and seven time types.

5.3.1 Group-level Analysis

We first conducted the group-level analysis and focused on the workload and performance aspects. We aimed at investigating Question 1, which is concerned with performance differences. For simplicity, we hereby refer to the five resource groups by short names, e.g., "muni-1" denotes the first municipality.

Workload analysis

We organized events and cases along the execution contexts to compare the workload of resource groups. Figure 5.4 shows the visualization of group workload in terms of cases organized by case types, and events organized by activity and type times. The five groups show similarities regarding the types of cases they processed (Figure 5.4a), as the majority leaned toward handling the construction-related applications, especially muni-1 and muni-5. They also exhibit very similar patterns in terms of assigning their group workload according to different types of activities (Figure 5.4b). Slight differences can be observed as neither muni-4 nor muni-5 has worked on activities of type 6. Also, employees from muni-2 and muni-5 seem to have committed to more workload in executing activities of type 8 ("01_HOOFD_8"). An interesting observation is concerned with the weekday pattern shown in Figure 5.4c. Observe that muni-1 differs from the others as it had only 12% of its total workload assigned on Wednesdays. In the meantime, muni-2, muni-3, and muni-5 seem to form another cohort as Fridays were their least busy day. This may be related to different arrangements of office hours in the groups.

Performance analysis

Figure 5.5 presents an overview of group performance measured by indicator amount-related productivity and time-related productivity for different year-quarters. For the analysis in this part, we based our observations on work profiles starting from 2012 Q1, since we only included cases started after 2010-12-31 in our evaluation. Hence, the numbers related to case completion in the early quarters of 2011 do not reflect the actual performance (note that the mean case cycle time in the dataset is 91.1 days).

5.3 Case Study: One Process, Five Municipalities

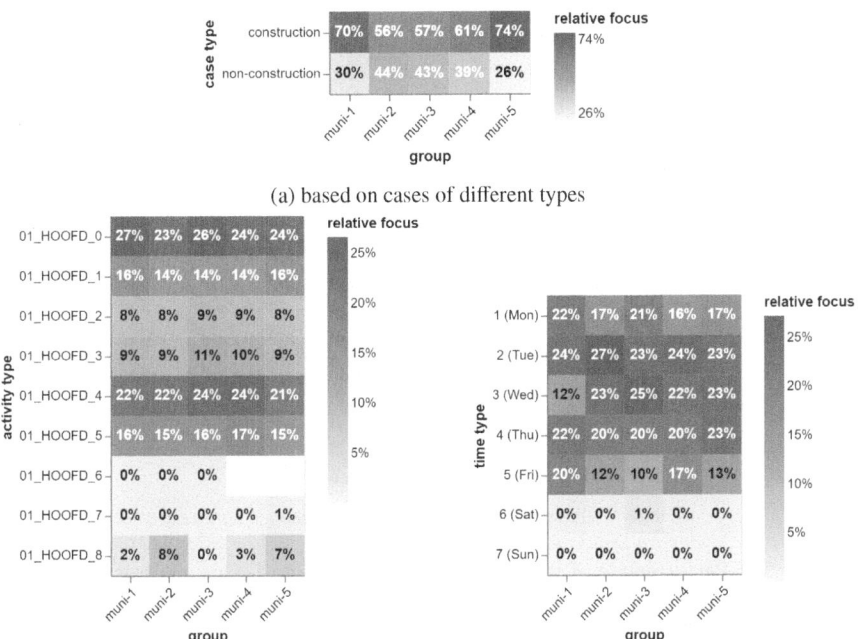

Fig. 5.4: Workload of the five groups in 2011–2014, measured by relative focus. The number "0%" corresponds to a rounded percentage value within the range $(0, 0.5\%)$, whereas a cell without annotation corresponds to a value of 0. Notice the similarities between the five groups regarding case types and activity types, and the differences regarding time types.

From Figure 5.5a, we can see that 2012 has the most completed cases. The groups' performance decreased in 2013 and went slightly higher in 2014. An observation worthwhile mentioning is that muni-4 had a sudden increase in performance in 2013 Q2 and 2013 Q4, and later decreased to a level comparable to the other groups. Figure 5.5b provides another perspective on group performance visualizing time-related productivity. Note that it is calculated by the average cycle time of completed cases, hence the performance is high when the value is low, and vice versa. We can see that muni-3 delivered steadily high performance in terms of shorter cycle time; muni-5 had a relatively consistent level of performance, which slightly improved during the year 2013; muni-1 and muni-4 were similar in general, except when approaching the end of 2014. Specifically, muni-2 stands out as its performance changed across the four quarters — within each year it started low in Q1, improved in Q2, and gradually decreased toward the end of the year (Q3 and Q4). This unique pattern of muni-2 would be of interest for further investigation.

Meanwhile, the spike in case cycle time in muni-1 and muni-4 in 2013 also deserves attention. With our previous observation on the increase of throughput of

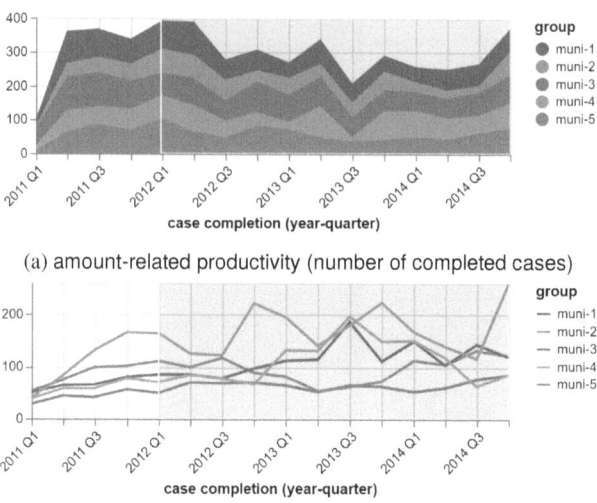

Fig. 5.5: Performance of the five groups in 2011–2014. Our analysis was based on data collected after 2011.

muni-4 in the same period, we selected the interval of 2013–2014 and used the detailed view to drill down on the performance of muni-4.

Figure 5.6 depicts the visualization. The upper view clearly shows four sharp increases of amount-related productivity. In each of the four weeks, muni-4 completed significantly more cases (more than 30) compared to all other groups (less than 10). This explains the spike in the overview (Figure 5.5a) and may suggest the existence of batching behavior of muni-4. Interestingly, the increase of amount-related productivity seems unrelated to the group's time-related productivity as shown in the lower view. Cross-checking the same weeks in the two charts, we can see that the potential batching completion did not directly link to a significantly longer case cycle time of muni-4.

5.3.2 Within-Group Analysis

We proceed to analysis at the group-member level motivated by Question 2, which considers the role differences between municipalities. Following the question, we analyzed the distribution within each group. We focused on the active group members, i.e., resources who committed to at least 1% of a group's activity allocation.

5.3 Case Study: One Process, Five Municipalities

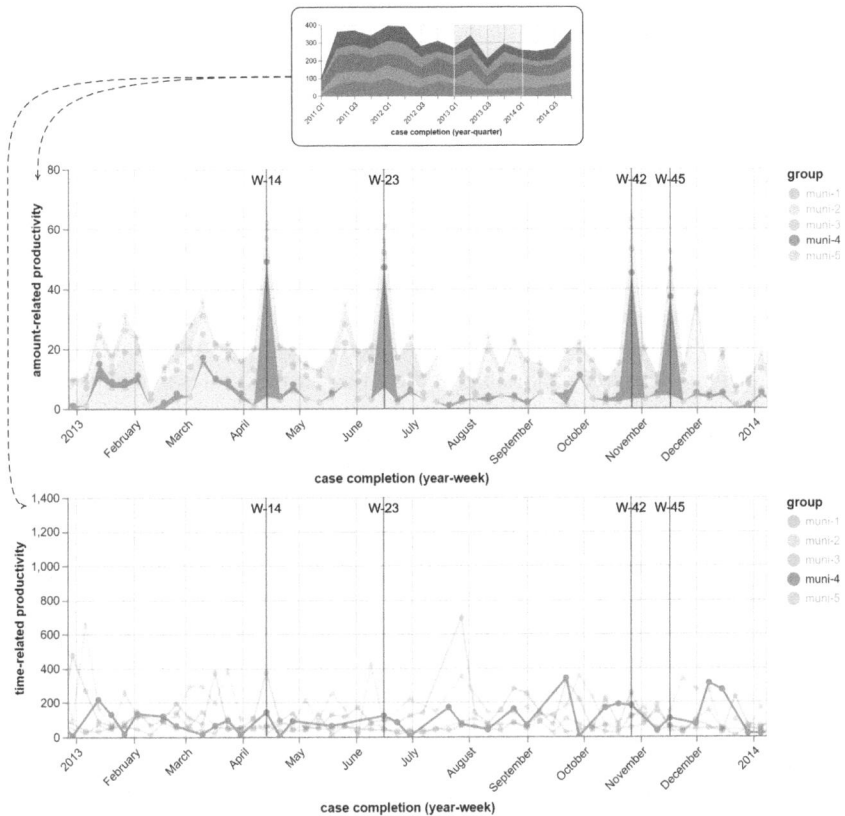

Fig. 5.6: Performance of muni-4 by amount-related productivity (upper chart) and time-related productivity (lower chart) in the selected interval 2013–2014. Notice the spikes in amount-related productivity — the vertical lines indicate the week numbers, e.g., "W-14" corresponds to the 14th week of the year.

Distribution analysis

Figure 5.7 presents how individual resources within each group handled events of different activity types and case types. The former reflects their participation at different phases ("01_HOOFD_0" to "01_HOOFD_8") of the permit application process, while the latter reflects their involvement in different categories of applications (construction vs. non-construction).

Comparing the columns in the heatmaps, we noticed two major cohorts within each of the five groups. This is the most significant in the cases of muni-4 and muni-5. On the one hand, there exists a cohort of resources focusing primarily on performing activities of type 0, 4, and 5, while they seldom carry out activities in the middle of the process (type 1, 2, and 3). Among them, there are a few that also showed similar distribution to construction- and non-construction-related cases (with either category taking 40% – 60% of an individual resource). On the other hand, there is a cohort of resources who were mostly executing activities from phases in the middle (types 1, 2, 3, and 4) in a balanced manner. This second cohort of resources was less involved in executing activities of types 0 and 5. Also, they were all relatively specialized in terms of handling the two types of cases (events of one case type dominating the other).

Let us take muni-4 as an example. (i) Resource "560752", "560781", and "560821" are the ones corresponding to the first cohort. In particular, "560752" and "560781" were distributed a comparable number of events from construction and non-construction cases (64% vs. 36%; 55% vs. 45%). (ii) All other members of the group correspond to the second cohort. These two different yet possibly complementary resource cohorts may reflect two business roles in the process.

The heatmaps also highlight patterns unique to some municipalities. For example, resource "560925" in muni-1 carried over 89% of its total workload in executing activities of type 0, and 8% in conducting activities of type 1. The resource was rarely involved in activities during the later phases of the process. While such a pattern is not observed in the other groups, it implies that muni-1 might have set up a specific role for dealing with the initial processing of the received applications.

As another example, resource "8492512" in muni-5 only executed activities of type 0, 4, and 5 in the four-year period, and may have acted as a specialist for the first major role identified previously (i.e., resource cohort focusing mainly on activities of types 0, 4, and 5). Similarly, resource "560752" may have served in the same specialist role in muni-5 — note that this resource had played the first major role when working for muni-4.

Lastly, in muni-4 and muni-5, there exist two resources that never executed activities in the context of construction-related cases ("560812" in muni-4 and "560596" in muni-5) and show patterns of the second major role (i.e., focusing mainly on the middle phases of the process). They may be staff who did not possess the required knowledge of permissions to deal with construction-related applications.

5.3.3 Summary

The above analyses of group work profiles using visual analytics revealed interesting patterns in terms of how five different resource groups worked on the same process. To address the first business question (throughput time difference), we analyzed the performance aspect measured by two indicators, amount-related productivity and time-related productivity. We concluded that group performance varies regarding time-related productivity. In particular, we found similarities between certain groups, i.e., muni-3 and muni-5, muni-1 and muni-4, and highlighted the specific yearly performance pattern exhibited by muni-2. We also compared the groups in terms of the workload over the four-year period, measured by relative focus. The differences between the groups mainly lie in the time dimension. While these observations cannot be used to directly explain the throughput time difference, they are useful insights for further investigation.

For the second question (role differences), we analyzed the distribution aspect based on the indicator member assignment. The analysis revealed two major roles, which focused on different phases of the process, and also identified resources and possible roles that were unique to certain groups.

5.4 Discussion

This chapter presents the notion of resource group work profiles — a collection of quantitative indicators measuring resource group performance in process execution from six aspects, which we synthesized from reviewing the management literature. Based on this notion, we introduced an approach to extracting and analyzing resource group profiles. First, given an event log and an organizational model (alternatively, domain knowledge specifying resource groupings and execution contexts), the pre-defined indicators can be calculated; then, visual analytics can be applied to the profiles to track, compare, and correlate different aspects of resource groups' performance. We tested our approach on a real-life event log dataset. The results reveal insightful patterns that can be used to answer questions related to workforce analytics proposed by the process owner. For the more complicated questions, our results can be used to pinpoint groups and aspects that require further investigation. While we did not aim at a thorough study of the resource groups in the dataset, we demonstrated that our approach can be applied to organizational models and event logs and supports group-oriented workforce analytics.

Our work has several contributions. First, the work profile indicators systematically extend the set of *model analysis* measures in the *OrdinoR* framework to more aspects relevant to workforce analytics, taking into account temporal changes. They broaden the way of diagnosing organizational models, e.g., one can find out exactly when a resource group has a low relative stake regarding an execution context, by checking the group participation and distribution at different time intervals. As such, it becomes possible to explain not just where an organizational model disagrees with an event log, but also to locate where the log disagrees with the model. Second, the proposed approach based on work profiles contributes an application of the *OrdinoR* framework. It enables the use of organizational models and event logs to objectively characterize and evaluate different employee groups over time. From a practical perspective, this provides organizations with the capability of continually adapting the organizational structures deployed around employees. Last but not least, our work also contributes to the use of visual analytics in process mining research.

As with any research, our work is subject to limitations. First, in terms of research methods, the synthesis of the work profile indicators will benefit from (i) conducting a systematic literature review of the dedicated literature to identify a more comprehensive collection of aspects and indicators, and (ii) using a requirement analysis to identify data attributes that need to be recorded by event logs concerning the extended aspects, or data sources in addition to event logs. Second, other methods for analyzing resource group work profiles should be explored. Visual analytics is a suitable means for descriptive analyses, especially during the initial exploration of the extracted profiles. But to explain the causes of observed issues or to make predictions for planning purposes, methods like correlation analysis and regression analysis should be applied.

5.4 Discussion

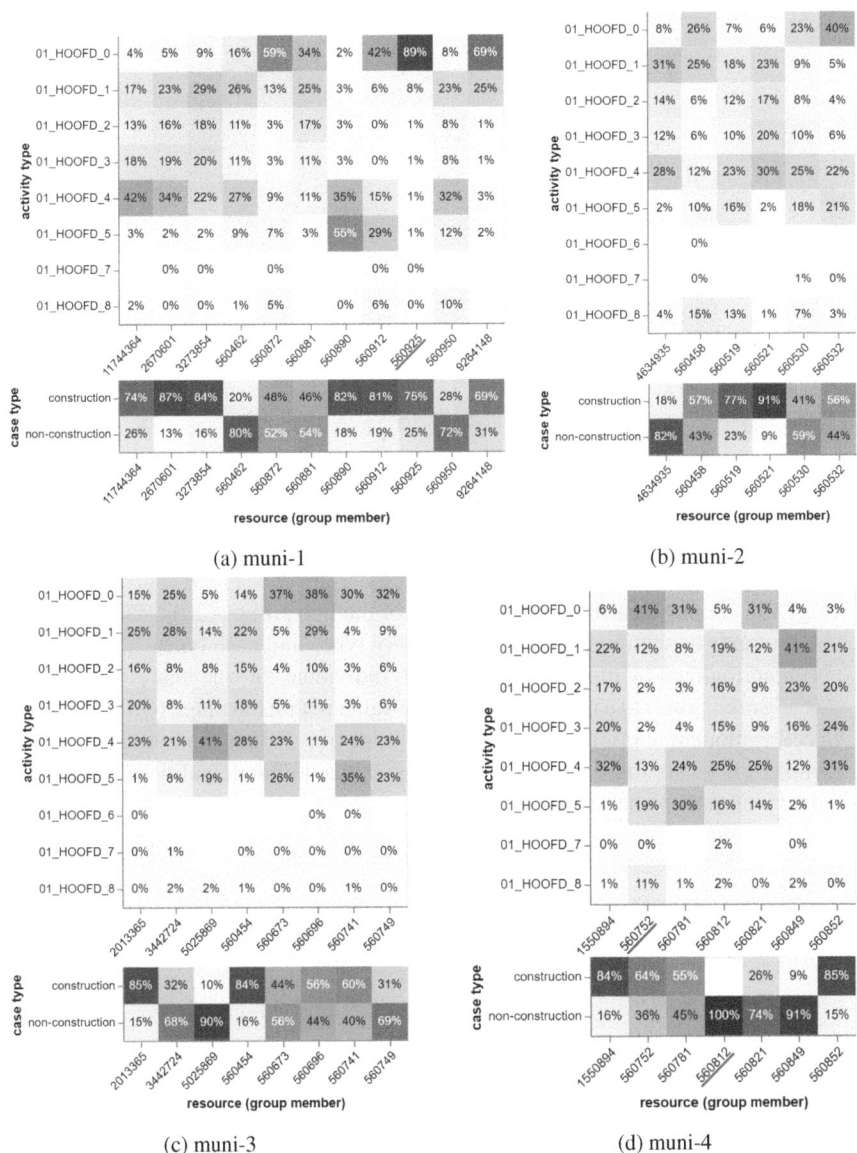

Fig. 5.7: Distribution within each of the five groups (2011–2014), measured by member assignment in terms of activity types and case types. The values have been normalized by member load of each individual for role analysis. The number "0%" corresponds to a rounded percentage value within the range $(0, 0.5\%)$, whereas a cell without annotation corresponds to a value of 0. Notice that resources annotated with red lines are those exhibiting patterns unique to municipalities, as discussed in the within-group analysis.

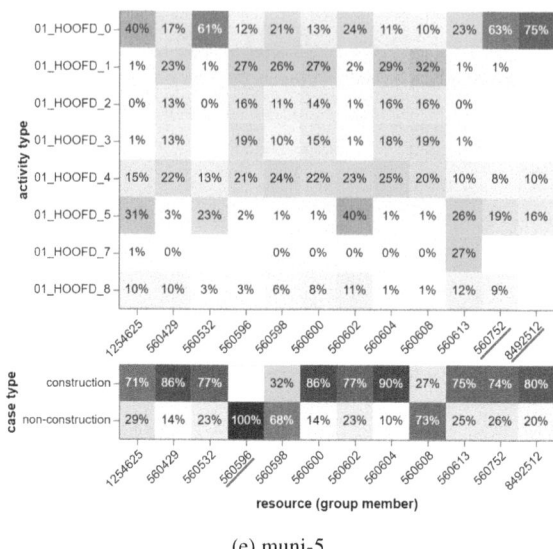

(e) muni-5

Fig. 5.7: (Cont.) Distribution within each of the five groups (2011–2014), measured by member assignment in terms of activity types and case types. The values have been normalized by member load of each individual for role analysis. The number "0%" corresponds to a rounded percentage value within the range $(0, 0.5\%)$, whereas a cell without annotation corresponds to a value of 0. Notice that resources annotated with red lines are those exhibiting patterns unique to municipalities, as discussed in the within-group analysis.

Chapter 6
Epilogue

> *Take history as a mirror, so in light of the past one can evaluate the present and see future trends. Take people and their words as a mirror, so one can reflect on gains and losses.*
>
> Taizong of the Tang empire

6.1 Conclusions

This research set out to explore process mining as a solution to extract knowledge about organizational groupings from event logs and use that knowledge to facilitate the management of human resources groups. Motivated by the research problem, we first conducted a thorough review of the existing organizational model mining approaches, which is a set of process mining techniques dedicated to discovering knowledge about resource groupings. Our review identified several key research gaps to be addressed. Therefore, we proposed a conceptual framework (*OrdinoR*) built around a novel definition of organizational model that describes both the grouping of resources and groups' involvement in process execution — the latter is neglected by many existing solutions. We also introduced fitness and precision as measures for evaluating the quality of organizational models, and a set of measures for analyzing organizational models. Guided by the framework, we proposed an approach to discovering organizational models. It starts with the learning of execution contexts, which utilizes the so-called type-defining attributes recorded in event logs to derive a set of logical rules that best describe the specialization of resources in process execution. The learned execution contexts can then be used to identify the grouping of resources sharing similar characteristics and linked to the groups to describe their involvement in process execution. As a result, organizational models are discovered from the input event logs. Experiments were conducted using real-life event logs collected from processes of three business domains. The results demonstrated that the proposed approach is feasible and effective. We also explored the use of organizational models for group-oriented workforce analytics. We introduced the notion of resource group work profile, which can measure six aspects of how resource groups work in process execution using various indicators, across group and individual levels, over different time periods, and along multiple process dimensions. We also proposed an approach to extract these work profiles from event logs and analyze them using visual analytics. We conducted a case study on an event log dataset that records five resource groups performing the same process, which demonstrated the usefulness of using work profiles for analyses of resource groups and their members in the context of process execution.

This doctoral research contributes to the field of process mining from the organizational perspective [7]. Specifically, it addressed three research gaps concerned with organizational model mining by (i) utilizing multidimensional event log information for model discovery, (ii) improving the interpretability of organizational models by capturing resource group involvement in process execution, and (iii) enabling a generic method for assessing the quality of discovered models with model evaluation and model analysis measures. Furthermore, our approach to extracting and analyzing resource group work profiles presents a novel means to exploit discovered organizational models for analyzing resources and their groupings. This enhances the practical use of organizational model mining. Last but not least, the notion of execution contexts and the approach to learning them from event logs also contribute to other research topics of process mining from the organizational perspective, e.g., analyzing individual resources and comparing them along different process dimensions.

Our work also contributes to the field of human resource management on the topic of workforce analytics. We introduced event logs as a useful data source and how they can be exploited using our approaches. Our experiment results showcased the feasibility of using event logs for the analysis of group workload, performance, and work distribution over group members of various roles. In doing so, our research may be of assistance to the discussion of connecting human resource management to the domain of business process management [75].

6.2 Future Work

Our work paves many future avenues for both research and application. We discuss some of them below.

Conformance checking of organizational models

This is concerned with comparing modeled behavior with the observed, real behavior recorded in event logs [7]. To begin with, organizational models are constructed from existing employee groupings, such as departments, business roles, and project teams. This can be done by identifying (i) employee groups and their members involved in process execution, and (ii) the patterns or rules regarding how process activities are performed based on the grouping, in order to specify types for defining execution contexts. Data other than event logs may be needed, e.g., documents about work distribution rules and timetables showing employee shifts. Then, given an organizational model, either discovered or based on existing employee groups, conformance checking can be performed. More specifically, fitness and precision in the *OrdinoR* framework can be used for *global conformance checking* — measuring the extent of commonalities between the modeled and actual behavior of human resource groups. The model analysis measures, e.g., group relative stake, can be used for *local conformance checking* — showing where and how the modeled human resource groups differ from reality. Hence, by combining conformance checking with organizational model mining capability, the *OrdinoR* framework lays the foundation for systematically exploiting process execution data to guide the design of organizational structures (covering, e.g., role designation and employee team composition) and staff deployment alongside changing business processes. Organizations can thus be empowered to evolve organizational structures toward process improvement, and to iteratively evaluate their decisions to enhance coordination, team effectiveness, and work satisfaction. To support this capacity-building, future work can investigate additional information needed in event logs to measure those goals.

Introducing additional quality evaluation measures

In the *OrdinoR* framework, two measures are proposed for evaluating the quality of discovered organizational models, i.e., fitness, which measures model completeness, and precision, which measures model exactness. More quality dimensions can be considered and will require additional evaluation measures. For instance, when two organizational models discovered from the same event log have the same level of fitness and precision, it will be interesting to consider the extent of model *simplicity*. A simpler model that can explain the same behavior observed in event log data may be preferred. Also, from the perspective of knowledge discovery from data, it is important to consider the extent of *generalization* of a discovered model, i.e., the ability of the model to describe resource groupings based on observations captured by multiple event logs of the same process, instead of overfitting just the examples in the input event log.

Extension to the organizational model definition

The current definition of organizational models (Definition 2.5) captures the grouping of resources and their involvement in process execution via execution contexts. Note that from an organizational structure point of view [29], such a resource grouping is "flat". Therefore, an extension to the model definition can be to consider hierarchical relations among resource groups. It would also be interesting to consider extending the set of "core event attributes" for execution contexts, so that organizational models can capture additional process dimensions, such as geographical locations, staff costs, etc., when such information is recorded in event logs. Of course, these extensions will trigger new challenges in terms of the approaches to discovering models.

Application of organizational models to process simulation

Simulation constitutes an important feature of many business process management tools. Having an approach that extends state-of-the-art process simulation techniques with organizational models can help validate potential modifications of an organizational model. Solutions to this issue will complement the approaches presented in this book and provide answers to "what-if" questions in workforce analytics, further enhancing the support for evaluating alternative decisions on organizational resource management.

Responsible use of organizational model mining

This research is concerned data-driven approaches supporting human-related decisions. Therefore, it is essential for future research to explore privacy and ethics concerns around the application of the approaches and ensure legitimate and responsible usage.

Future development of process mining software tools

Last but not the least, outcomes from this research may contribute to informing future development of process mining software and workforce analytics software. Analysis of a business process can be systematically linked with the analysis of human resources and groups through the discovery of organizational models and the subsequent resource group work profiles. In addition, the visual analytics design presented in this research may inform the development of more powerful dashboard tools for better understanding employees along multiple dimensions.

References

1. Gartner Magic Quadrant for Process Mining Platforms. https://www.gartner.com/en/documents/5391263. Accessed: 2025-4-3
2. Gartner Research – Market Guide for Process Mining. https://www.gartner.com/en/documents/3939836. Accessed: 2023-3-10
3. Organizational extension - IEEE Task Force on Process Mining. https://www.tf-pm.org/resources/xes-standard/about-xes/standard-extensions/org. Accessed: 2023-6-22
4. IEEE Standard for eXtensible Event Stream (XES) for Achieving Interoperability in Event Logs and Event Streams. IEEE Std 1849-2016 pp. 1–50 (2016). DOI 10.1109/IEEESTD.2016.7740858. URL http://dx.doi.org/10.1109/IEEESTD.2016.7740858
5. van der Aalst, W.M.P.: Process Mining: Overview and Opportunities. ACM Transactions on Management Information Systems **3**(2), 1–17 (2012).
6. van der Aalst, W.M.P.: Process Cubes: Slicing, Dicing, Rolling Up and Drilling Down Event Data for Process Mining. In: M. Song, M.T. Wynn, J. Liu (eds.) Asia Pacific Business Process Management - First Asia Pacific Conference, AP-BPM 2013, Beijing, China, August 29-30, 2013. Selected Papers, Lecture Notes in Business Information Processing, pp. 1–22. Springer, Cham (2013).
7. van der Aalst, W.M.P.: Process Mining: Data Science in Action. Springer (2016). DOI 10.1007/978-3-662-49851-4_1. URL https://play.google.com/store/books/details?id=hUEGDAAAQBAJ
8. van der Aalst, W.M.P., Carmona, J. (eds.): Process Mining Handbook, 1 edn. Lecture notes in business information processing. Springer International Publishing, Cham, Switzerland (2022). DOI 10.1007/978-3-031-08848-3. URL https://link.springer.com/book/10.1007/978-3-031-08848-3
9. van der Aalst, W.M.P., Reijers, H.A., Song, M.: Discovering Social Networks from Event Logs. Computer Supported Cooperative Work **14**(6), 549–593 (2005).
10. van der Aalst, W.M.P., Song, M.: Mining Social Networks: Uncovering Interaction Patterns in Business Processes. In: J. Desel, B. Pernici, M. Weske (eds.) Business Process Management: Second International Conference, BPM 2004, Potsdam, Germany, June 17-18, 2004. Proceedings, pp. 244–260. Springer, Berlin, Heidelberg (2004)
11. Aarts, E., Korst, J., Michiels, W.: Simulated Annealing. In: E.K. Burke, G. Kendall (eds.) Search Methodologies: Introductory Tutorials in Optimization and Decision Support Techniques, pp. 187–210. Springer US, Boston, MA (2005).
12. Acheli, M., Grigori, D., Weidlich, M.: Discovering and Analyzing Contextual Behavioral Patterns From Event Logs. IEEE Transactions on Knowledge and Data Engineering **34**(12), 5708–5721 (2022). DOI 10.1109/TKDE.2021.3077653. URL http://dx.doi.org/10.1109/TKDE.2021.3077653
13. Appice, A.: Towards mining the organizational structure of a dynamic event scenario. Journal of Intelligent Information Systems **50**(1), 165–193 (2018).
14. Arthur, D., Vassilvitskii, S.: k-means++: the advantages of careful seeding. In: Proceedings of the Eighteenth Annual ACM-SIAM Symposium on Discrete Algorithms, SODA 2007, New Orleans, Louisiana, USA, January 7-9, 2007, SODA '07, pp. 1027–1035. Society for Industrial and Applied Mathematics, USA (2007). URL https://dl.acm.org/doi/abs/10.5555/1283383.1283494

15. Banerjee, A., Krumpelman, C., Ghosh, J., Basu, S., Mooney, R.J.: Model-based overlapping clustering. In: Proceedings of the Eleventh ACM SIGKDD International Conference on Knowledge Discovery and Data Mining, Chicago, Illinois, USA, August 21-24, 2005, KDD '05, pp. 532–537. Association for Computing Machinery, New York, NY, USA (2005).
16. Baumgrass, A.: Deriving Current State RBAC Models from Event Logs. In: Sixth International Conference on Availability, Reliability and Security, ARES 2011, Vienna, Austria, August 22-26, 2011, pp. 667–672. IEEE Computer Society, Vienna (2011). DOI 10.1109/ARES.2011.104. URL http://dx.doi.org/10.1109/ARES.2011.104
17. Baumgrass, A., Baier, T., Mendling, J., Strembeck, M.: Conformance Checking of RBAC Policies in Process-Aware Information Systems. In: F. Daniel, K. Barkaoui, S. Dustdar (eds.) Business Process Management Workshops - BPM 2011 International Workshops, Clermont-Ferrand, France, August 29, 2011, Revised Selected Papers, Part II, pp. 435–446. Springer, Berlin, Heidelberg (2012). DOI 10.1007/978-3-642-28115-0_41. URL http://dx.doi.org/10.1007/978-3-642-28115-0_41
18. Berti, A., van Zelst, S., Schuster, D.: PM4Py: A process mining library for Python. Software Impacts **17**, 100556 (2023). DOI 10.1016/j.simpa.2023.100556. URL https://www.sciencedirect.com/science/article/pii/S2665963823000933
19. Bolt, A., van der Aalst, W.M.P.: Multidimensional Process Mining Using Process Cubes. In: K. Gaaloul, R. Schmidt, S. Nurcan, S. Guerreiro, Q. Ma (eds.) Enterprise, Business-Process and Information Systems Modeling - 16th International Conference, BPMDS 2015, 20th International Conference, EMMSAD 2015, Held at CAiSE 2015, Stockholm, Sweden, June 8-9, 2015, pp. 102–116. Springer International Publishing (2015). DOI 10.1007/978-3-319-19237-6_7. URL http://dx.doi.org/10.1007/978-3-319-19237-6_7
20. Bortoluzzi, B., Carey, D., McArthur, J.J., Menassa, C.: Measurements of workplace productivity in the office context: A systematic review and current industry insights. Journal of Corporate Real Estate **20**(4), 281–301 (2018).
21. Bose, R.P.J.C., van der Aalst, W.M.P.: Context aware trace clustering: Towards improving process mining results. In: Proceedings of the SIAM International Conference on Data Mining, SDM 2009, April 30 - May 2, 2009, Sparks, Nevada, USA, pp. 401–412. Society for Industrial and Applied Mathematics, Philadelphia, PA (2009). DOI 10.1137/1.9781611972795.35. URL https://epubs.siam.org/doi/10.1137/1.9781611972795.35
22. Bouzguenda, L., Abdelkafi, M.: An agent-based approach for organizational structures and interaction protocols mining in workflow. Social Network Analysis and Mining **5**(1), 10 (2015).
23. Brignall, S., Ballantine, J.: Performance measurement in service businesses revisited. International Journal of Service Industry Management **7**(1), 6–31 (1996).
24. Buijs, J.C.A.M.: Flexible evolutionary algorithms for mining structured process models. Ph.D. thesis, Eindhoven University of Technology (2014)
25. Buijs, J.C.A.M.: Receipt phase of an environmental permit application process (WABO), CoSeLoG project (2022). DOI 10.4121/12709127.V2. URL https://data.4tu.nl/articles/_/12709127/2
26. Burattin, A., Sperduti, A., Veluscek, M.: Business models enhancement through discovery of roles. In: IEEE Symposium on Computational Intelligence and Data Mining, CIDM 2013, Singapore, 16-19 April, 2013, pp. 103–110. IEEE (2013). DOI 10.1109/CIDM.2013.6597224. URL http://dx.doi.org/10.1109/CIDM.2013.6597224
27. Charlwood, A., Stuart, M., Trusson, C.: Human capital metrics and analytics: assessing the evidence of the value and impact of people data. Tech. rep. (2017)
28. Coelli, T.J., Prasada Rao, D.S., O'Donnell, C.J., Battese, G.E.: An Introduction to Efficiency and Productivity Analysis. Springer US (2005). DOI 10.1007/b136381. URL https://link.springer.com/book/10.1007/b136381

References

29. Daft, R.L., Murphy, J., Willmott, H.: Organization theory and design. Cengage learning EMEA (2010). URL https://ebookcentral.proquest.com/lib/qut/detail.action?docID=4635701
30. Davenport, T.H., Harris, J., Shapiro, J.: Competing on talent analytics. Harvard Business Review **88**(10), 52–58 (2010). URL https://www.ncbi.nlm.nih.gov/pubmed/20929194
31. Davenport, T.H., Spanyi, A.: What Process Mining Is, and Why Companies Should Do It. Harvard Business Review (2019). URL https://hbr.org/2019/04/what-process-mining-is-and-why-companies-should-do-it
32. Delcoucq, L., Lecron, F., Fortemps, P., van der Aalst, W.M.P.: Resource-centric process mining: clustering using local process models. In: C.C. Hung, T. Cerny, D. Shin, A. Bechini (eds.) SAC '20: The 35th ACM/SIGAPP Symposium on Applied Computing, online event, [Brno, Czech Republic], March 30 - April 3, 2020, SAC '20, pp. 45–52. Association for Computing Machinery, New York, NY, USA (2020). DOI 10.1145/3341105.3373864. URL https://doi.org/10.1145/3341105.3373864
33. van Dongen, B.F.: BPI Challenge 2015 (2015). DOI 10.4121/UUID:31A308EF-C844-48DA-948C-305D167A0EC1. URL https://data.4tu.nl/collections/BPI_Challenge_2015/5065424
34. van Dongen, B.F.: BPI Challenge 2017 (2017). DOI 10.4121/UUID:5F3067DF-F10B-45DA-B98B-86AE4C7A310B. URL https://data.4tu.nl/articles/dataset/BPI_Challenge_2017/12696884
35. van Dongen, B.F., Borchert, F.: BPI Challenge 2018 (2018). DOI 10.4121/UUID:3301445F-95E8-4FF0-98A4-901F1F204972. URL https://data.4tu.nl/articles/dataset/BPI_Challenge_2018/12688355
36. Dumas, M., van der Aalst, W.M.P., ter Hofstede, A.H.M.: Process-Aware Information Systems: Bridging People and Software Through Process Technology. Wiley (2005). DOI 10.1002/0471741442. URL https://play.google.com/store/books/details?id=nZ5QAAAAMAAJ
37. Dumas, M., La Rosa, M., Mendling, J., Reijers, H.A.: Fundamentals of Business Process Management. Springer (2018). DOI 10.1007/978-3-662-56509-4. URL https://play.google.com/store/books/details?id=VQk4tAEACAAJ
38. van Eck, M.L., Lu, X., Leemans, S.J.J., van der Aalst, W.M.P.: PM^2: A Process Mining Project Methodology. In: J. Zdravkovic, M. Kirikova, P. Johannesson (eds.) Advanced Information Systems Engineering - 27th International Conference, CAiSE 2015, Stockholm, Sweden, June 8-12, 2015, Proceedings, pp. 297–313. Springer (2015). DOI 10.1007/978-3-319-19069-3_19. URL http://dx.doi.org/10.1007/978-3-319-19069-3_19
39. Ferreira, D.R., Alves, C.: Discovering User Communities in Large Event Logs. In: F. Daniel, K. Barkaoui, S. Dustdar (eds.) Business Process Management Workshops - BPM 2011 International Workshops, Clermont-Ferrand, France, August 29, 2011, Revised Selected Papers, Part I, pp. 123–134. Springer, Berlin, Heidelberg (2012). DOI 10.1007/978-3-642-28108-2_11. URL http://dx.doi.org/10.1007/978-3-642-28108-2_11
40. Garvin, D.A.: How Google sold its engineers on management. Harvard Business Review **91**(12), 74–82 (2013)
41. Gibson, C.B., Zellmer-Bruhn, M.E., Schwab, D.P.: Team Effectiveness in Multinational Organizations: Evaluation Across Contexts. Group & Organization Management **28**(4), 444–474 (2003).
42. Han, J., Kamber, M., Pei, J.: Data Mining: Concepts and Techniques. Elsevier (2011). URL https://play.google.com/store/books/details?id=pQws07tdpjoC
43. Hanachi, C., Gaaloul, W., Mondi, R.: Performative-Based Mining of Workflow Organizational Structures. In: C. Huemer, P. Lops (eds.) Proceedings of the 13th International Conference on E-Commerce and Web Technologies (EC-Web 2012), pp. 63–75. Springer, Berlin, Heidelberg (2012)
44. Harris, J.G., Craig, E., Light, D.A.: Talent and analytics: new approaches, higher ROI. Journal of Business Strategy **32**(6), 4–13 (2011).

45. Haynes, B.P.: An evaluation of office productivity measurement. Journal of Corporate Real Estate **9**(3), 144–155 (2007).
46. Henderson, D., Jacobson, S.H., Johnson, A.W.: The Theory and Practice of Simulated Annealing. In: F. Glover, G.A. Kochenberger (eds.) Handbook of Metaheuristics, pp. 287–319. Springer US, Boston, MA (2003).
47. van den Heuvel, S., Bondarouk, T.: The rise (and fall?) of HR analytics: A study into the future application, value, structure, and system support. Journal of Organizational Effectiveness **4**(2), 157–178 (2017). DOI 10.1108/JOEPP-03-2017-0022. URL http://dx.doi.org/10.1108/JOEPP-03-2017-0022
48. Huang, Z., Lu, X., Duan, H.: Mining association rules to support resource allocation in business process management. Expert Systems with Applications **38**(8), 9483–9490 (2011). DOI 10.1016/j.eswa.2011.01.146. URL https://www.sciencedirect.com/science/article/pii/S0957417411001795
49. Huang, Z., Lu, X., Duan, H.: Resource behavior measure and application in business process management. Expert Systems with Applications **39**(7), 6458–6468 (2012). DOI 10.1016/j.eswa.2011.12.061. URL https://www.sciencedirect.com/science/article/pii/S0957417411017325
50. van Hulzen, G., Martin, N., Depaire, B.: Looking Beyond Activity Labels: Mining Context-Aware Resource Profiles Using Activity Instance Archetypes. In: A. Polyvyanyy, M.T. Wynn, A. Van Looy, M. Reichert (eds.) Business Process Management Forum - BPM Forum 2021, Rome, Italy, September 06-10, 2021, Proceedings, pp. 230–245. Springer (2021). DOI 10.1007/978-3-030-85440-9_14. URL http://dx.doi.org/10.1007/978-3-030-85440-9_14
51. Jin, T., Wang, J., Wen, L.: Organizational modeling from event logs. In: Y. Han, G. Alonso, R. Buyya, C. Xu (eds.) Sixth International Conference on Grid and Cooperative Computing (GCC 2007), 16–18 August 2007, Urumchi, Xinjiang, China, pp. 670–675. IEEE Computer Society (2007). DOI 10.1109/GCC.2007.93. URL http://dx.doi.org/10.1109/GCC.2007.93
52. Keim, D.A., Mansmann, F., Schneidewind, J., Thomas, J., Ziegler, H.: Visual Analytics: Scope and Challenges. In: S.J. Simoff, M.H. Böhlen, A. Mazeika (eds.) Visual Data Mining: Theory, Techniques and Tools for Visual Analytics, vol. 4404 LNCS, pp. 76–90. Springer (2008). DOI 10.1007/978-3-540-71080-6_6. URL http://infovis.uni-konstanz.de
53. Kirkpatrick, S., Gelatt Jr, C.D., Vecchi, M.P.: Optimization by simulated annealing. Science **220**(4598), 671–680 (1983). DOI 10.1126/science.220.4598.671. URL http://dx.doi.org/10.1126/science.220.4598.671
54. Kohavi, R., Li, C.H.: Oblivious Decision Trees Graphs and Top down Pruning. In: Proceedings of the 14th International Joint Conference on Artificial Intelligence - Volume 2, IJCAI'95, pp. 1071–1077. Morgan Kaufmann Publishers Inc., San Francisco, CA, USA (1995). URL https://dl.acm.org/doi/abs/10.5555/1643031.1643039
55. Kumar, A., Liu, S.: Analyzing a Helpdesk Process Through the Lens of Actor Hand-off Patterns. In: D. Fahland, C. Ghidini, J. Becker, M. Dumas (eds.) Business Process Management Forum - BPM Forum 2020, Seville, Spain, September 13-18, 2020, Proceedings, Lecture notes in business information processing, pp. 313–329. Springer International Publishing, Cham (2020). DOI 10.1007/978-3-030-58638-6_19. URL https://link.springer.com/10.1007/978-3-030-58638-6_19
56. Levenson, A.: Using workforce analytics to improve strategy execution. Human Resource Management **57**(3), 685–700 (2018). DOI 10.1002/hrm.21850. URL https://onlinelibrary.wiley.com/doi/10.1002/hrm.21850
57. Li, M., Liu, L., Yin, L., Zhu, Y.: A process mining based approach to knowledge maintenance. Information Systems Frontiers **13**(3), 371–380 (2011).
58. Liu, R., Agarwal, S., Sindhgatta, R.R., Lee, J.: Accelerating Collaboration in Task Assignment Using a Socially Enhanced Resource Model. In: F. Daniel, J. Wang, B. Weber (eds.) Business

Process Management: 11th International Conference, BPM 2013, Beijing, China, August 26–30, 2013. Proceedings, pp. 251–258. Springer, Berlin, Heidelberg (2013). DOI 10.1007/978-3-642-40176-3_21. URL http://dx.doi.org/10.1007/978-3-642-40176-3_21

59. Liu, T., Cheng, Y., Ni, Z.: Mining event logs to support workflow resource allocation. Knowledge-Based Systems **35**, 320–331 (2012). DOI 10.1016/j.knosys.2012.05.010. URL http://www.sciencedirect.com/science/article/pii/S0950705112001542
60. Liu, Y., Wang, J., Yang, Y., Sun, J.: A semi-automatic approach for workflow staff assignment. Computers in Industry **59**(5), 463–476 (2008). DOI 10.1016/j.compind.2007.12.002. URL http://dx.doi.org/10.1016/j.compind.2007.12.002
61. Lloyd, S.: Least squares quantization in PCM. IEEE Transactions on Information Theory **28**(2), 129–137 (1982). DOI 10.1109/TIT.1982.1056489. URL http://dx.doi.org/10.1109/TIT.1982.1056489
62. Ly, L.T., Rinderle, S., Dadam, P., Reichert, M.: Mining Staff Assignment Rules from Event-Based Data. In: C.J. Bussler, A. Haller (eds.) Business Process Management Workshops, BPM 2005 International Workshops, BPI, BPD, ENEI, BPRM, WSCOBPM, BPS, Nancy, France, September 5, 2005, Revised Selected Papers, pp. 177–190. Springer, Berlin, Heidelberg (2005).
63. Mannhardt, F.: Sepsis Cases - Event Log (2016). DOI 10.4121/UUID:915D2BFB-7E84-49AD-A286-DC35F063A460. URL https://data.4tu.nl/articles/_/12707639/1
64. Mannhardt, F., Blinde, D.: Analyzing the Trajectories of Patients with Sepsis using Process Mining. In: J. Gulden, S. Nurcan, I. Reinhartz-Berger, W. Guédria, P. Bera, S. Guerreiro, M. Fellmann, M. Weidlich (eds.) Joint Proceedings of the Radar tracks at the 18th International Working Conference on Business Process Modeling, Development and Support (BPMDS), and the 22nd International Working Conference on Evaluation and Modeling Methods for Systems Analysis and Development (EMMSAD), and the 8th International Workshop on Enterprise Modeling and Information Systems Architectures (EMISA) co-located with the 29th International Conference on Advanced Information Systems Engineering 2017 (CAiSE 2017), Essen, Germany, June 12-13, 2017, *CEUR Workshop Proceedings*, vol. 1859, pp. 72–80. CEUR-WS.org (2017). URL https://ceur-ws.org/Vol-1859/bpmds-08-paper.pdf
65. Marler, J.H., Boudreau, J.W.: An evidence-based review of HR Analytics. The International Journal of Human Resource Management **28**(1), 3–26 (2017). DOI 10.1080/09585192.2016.1244699. URL https://www.tandfonline.com/doi/full/10.1080/09585192.2016.1244699
66. Nakatumba, J., van der Aalst, W.M.P.: Analyzing Resource Behavior Using Process Mining. In: S. Rinderle-Ma, S. Sadiq, F. Leymann (eds.) Business Process Management Workshops: BPM 2009 International Workshops, Ulm, Germany, September 7, 2009. Revised Papers, pp. 69–80. Springer, Berlin, Heidelberg (2010).
67. Ni, Z., Wang, S., Li, H.: Mining organizational structure from workflow logs. In: Proceeding of the International Conference on e-Education, Entertainment and e-Management (ICeEEM 2011), pp. 222–225 (2011).
68. Pika, A., Leyer, M., Wynn, M.T., Fidge, C.J., ter Hofstede, A.H.M., van der Aalst, W.M.P.: Mining Resource Profiles from Event Logs. ACM Transactions on Management Information Systems **8**(1), 1–30 (2017).
69. Rinderle-Ma, S., van der Aalst, W.M.P.: Life-Cycle Support for Staff Assignment Rules in Process-Aware Information Systems. Tech. rep., Technische Universiteit Eindhoven (2007)
70. Rousseeuw, P.J.: Silhouettes: A graphical aid to the interpretation and validation of cluster analysis. Journal of Computational and Applied Mathematics **20**, 53–65 (1987). DOI 10.1016/0377-0427(87)90125-7. URL https://www.sciencedirect.com/science/article/pii/0377042787901257
71. Russell, N., van der Aalst, W.M.P., ter Hofstede, A.H.M.: Workflow Patterns: The Definitive Guide. The MIT Press (2016)

72. Satyanarayan, A., Moritz, D., Wongsuphasawat, K., Heer, J.: Vega-Lite: A Grammar of Interactive Graphics. IEEE Transactions on Visualization and Computer Graphics **23**(1), 341–350 (2017). DOI 10.1109/TVCG.2016.2599030. URL http://ieeexplore.ieee.org/document/7539624/
73. Schönig, S., Cabanillas, C., Jablonski, S., Mendling, J.: A framework for efficiently mining the organisational perspective of business processes. Decision Support Systems **89**, 87–97 (2016). DOI 10.1016/j.dss.2016.06.012. URL http://www.sciencedirect.com/science/article/pii/S016792361630104X
74. Sellami, R., Gaaloul, W., Moalla, S.: An Ontology for Workflow Organizational Model Mining. In: S. Reddy, K. Drira (eds.) 2012 IEEE 21st International Workshop on Enabling Technologies: Infrastructure for Collaborative Enterprises, pp. 199–204. IEEE (2012). DOI 10.1109/WETICE.2012.29. URL http://dx.doi.org/10.1109/WETICE.2012.29
75. Shafagatova, A., Van Looy, A.: Alignment patterns for process-oriented appraisals and rewards: using HRM for BPM capability building. Business Process Management Journal **27**(3), 941–964 (2020).
76. Smith, K.I., Everson, R.M., Fieldsend, J.E., Murphy, C., Misra, R.: Dominance-Based Multiobjective Simulated Annealing. IEEE Transactions on Evolutionary Computation **12**(3), 323–342 (2008). DOI 10.1109/TEVC.2007.904345. URL http://dx.doi.org/10.1109/TEVC.2007.904345
77. Song, M., van der Aalst, W.M.P.: Towards comprehensive support for organizational mining. Decision Support Systems **46**(1), 300–317 (2008). DOI 10.1016/j.dss.2008.07.002. URL http://dx.doi.org/10.1016/j.dss.2008.07.002
78. Song, M., Günther, C.W., van der Aalst, W.M.P.: Trace Clustering in Process Mining. In: D. Ardagna, M. Mecella, J. Yang (eds.) Business Process Management Workshops, BPM 2008 International Workshops, Milano, Italy, September 1-4, 2008. Revised Papers, pp. 109–120. Springer, Berlin, Heidelberg (2009). DOI 10.1007/978-3-642-00328-8_11. URL http://dx.doi.org/10.1007/978-3-642-00328-8_11
79. Suman, B., Kumar, P.: A survey of simulated annealing as a tool for single and multiobjective optimization. Journal of the Operational Research Society **57**(10), 1143–1160 (2006).
80. Suriadi, S., Wynn, M.T., Ouyang, C., ter Hofstede, A.H.M., van Dijk, N.J.: Understanding Process Behaviours in a Large Insurance Company in Australia: A Case Study. In: C. Salinesi, M.C. Norrie, O. Pastor (eds.) Advanced Information Systems Engineering - 25th International Conference, CAiSE 2013, Valencia, Spain, June 17-21, 2013. Proceedings, Lecture Notes in Computer Science, pp. 449–464. Springer (2013). DOI 10.1007/978-3-642-38709-8_29. URL http://dx.doi.org/10.1007/978-3-642-38709-8_29
81. Suriadi, S., Wynn, M.T., Xu, J., van der Aalst, W.M.P., ter Hofstede, A.H.M.: Discovering work prioritisation patterns from event logs. Decision Support Systems **100**, 77–92 (2017). DOI 10.1016/J.DSS.2017.02.002. URL https://www.sciencedirect.com/science/article/pii/S0167923617300180
82. Tan, P.N., Steinbach, M., Karpatne, A., Kumar, V.: Introduction to Data Mining. Pearson Education, Harlow, United Kingdom (2021). URL https://play.google.com/store/books/details?id=i8AoEAAAQBAJ
83. Tarique, I.: Seven Trends in Corporate Training and Development: Strategies to Align Goals with Employee Needs. Pearson Education (2014). URL https://play.google.com/store/books/details?id=yD5iAwAAQBAJ
84. Tax, N., Sidorova, N., Haakma, R., van der Aalst, W.M.P.: Mining local process models. Journal of Innovation in Digital Ecosystems **3**(2), 183–196 (2016). DOI 10.1016/j.jides.2016.11.001. URL http://www.sciencedirect.com/science/article/pii/S2352664516300232
85. Viner, D., Stierle, M., Matzner, M.: A process mining software comparison. In: C. Di Ciccio, B. Depaire, J. De Weerdt, C. Di Francescomarino, J. Munoz-Gama (eds.) Proceedings of the ICPM Doctoral Consortium and Tool Demonstration Track 2020 co-located with the

2nd International Conference on Process Mining (ICPM 2020). ceur-ws.org (2020). URL https://ceur-ws.org/Vol-2703/paperTD1.pdf
86. Ward, J.H.: Hierarchical Grouping to Optimize an Objective Function. Journal of the American Statistical Association **58**(301), 236–244 (1963). DOI 10.1080/01621459.1963.10500845. URL https://www.tandfonline.com/doi/abs/10.1080/01621459.1963.10500845
87. White, S.R.: Concepts of scale in simulated annealing. In: AIP Conference Proceedings - The Physics of VLSI 1-3, August 1984, Palo Alto, CA, USA, vol. 122, pp. 261–270. American Institute of Physics (1984). DOI 10.1063/1.34823. URL https://aip.scitation.org/doi/abs/10.1063/1.34823
88. Yang, J., Ouyang, C., Pan, M., Yu, Y., ter Hofstede, A.H.M.: Finding the "Liberos": Discover Organizational Models with Overlaps. In: M. Weske, M. Montali, I. Weber, J. vom Brocke (eds.) Business Process Management - 16th International Conference, BPM 2018, Sydney, NSW, Australia, September 9-14, 2018, Proceedings, Lecture Notes in Computer Science, pp. 339–355. Springer (2018). DOI 10.1007/978-3-319-98648-7_20. URL http://dx.doi.org/10.1007/978-3-319-98648-7_20
89. Ye, J.H., Li, Z.W., Yi, K., Al-Ahmari, A.: Mining Resource Community and Resource Role Network from Event Logs. IEEE Access **6**, 77685–77694 (2018). DOI 10.1109/ACCESS.2018.2883774. URL https://ieeexplore.ieee.org/document/8550643/
90. van Zelst, S.J., van Dongen, B.F., van der Aalst, W.M.P.: Online Discovery of Cooperative Structures in Business Processes. In: C. Debruyne, H. Panetto, R. Meersman, T. Dillon, E. Kühn, D. O'Sullivan, C.A. Ardagna (eds.) On the Move to Meaningful Internet Systems: OTM 2016 Conferences - Confederated International Conferences: CoopIS, C&TC, and ODBASE 2016, Rhodes, Greece, October 24-28, 2016, Proceedings, Lecture Notes in Computer Science, pp. 210–228. Springer, Cham (2016).
91. Zhao, W., Lin, Q., Shi, Y., Fang, X.: Mining the Role-Oriented Process Models Based on Genetic Algorithm. In: Y. Tan, Y. Shi, Z. Ji (eds.) Advances in Swarm Intelligence - Third International Conference, ICSI 2012, Shenzhen, China, June 17-20, 2012 Proceedings, Part I, Lecture Notes in Computer Science, pp. 398–405. Springer, Berlin, Heidelberg (2012)
92. Zhao, W., Zhao, X.: Process Mining from the Organizational Perspective. In: Z. Wen, T. Li (eds.) Foundations of Intelligent Systems: Proceedings of the Eighth International Conference on Intelligent Systems and Knowledge Engineering, Shenzhen, China, Nov 2013 (ISKE 2013), pp. 701–708. Springer, Berlin, Heidelberg (2014). DOI 10.1007/978-3-642-54924-3_66. URL http://dx.doi.org/10.1007/978-3-642-54924-3_66

The manufacturer's authorised representative in the EU is Springer Nature Customer Service Centre GmbH, Europaplatz 3, 69115 Heidelberg, Germany. If you have any concerns regarding our products, please contact ProductSafety@springernature.com

Printed and bound by CPI Group (UK) Ltd, Croydon, CR0 4YY

26/03/2026

02078973-0006